AF792449

Advances in Technology-Rich Science Education

Volume 4

Series Editor

Kent J. Crippen, University of Florida, Gainesville, FL, USA

Editorial Board

Len Annetta, Citadel, Charleston, SC, USA

Grant Cooper, Curtin University, Perth, WA, Australia

Brayan Diaz, Utah State University, Logan, Utah, UT, USA

Jonas Hallström, Linköping University, Linköping, Sweden

Marcus Kubsch, Freie Universität Berlin, Berlin, Berlin, Germany

Xiufeng Liu, University of Macau, Macau, Taipa, China

Karen Murcia, Curtin University, Perth, WA, Australia

Ross Nehm, Stony Brook University, Stony Brook, New York, NY, USA

Knut Neumann, Leibniz Institute for Science and Mathe, Kiel, Germany

Xiaoming Zhai, 105J Aderhold Hall, University of Georgia, Athens, GA, USA

This book series provides a comprehensive and authoritative overview of current research, trends, and innovations that leverage educational technology to improve science teaching and learning. Contributions from a global community of scholars, educators, policymakers, and leaders address formal, non-formal, and informal science education contexts.

It provides a platform for sharing research, practice, development, and policy, highlighting emerging innovations, and examining how technology influences engagement, learning, and workforce readiness. Additionally, it also addresses key challenges and opportunities related to access, equity, inclusion, and digital literacies while fostering critical perspectives and discussion about technology's evolving role in science education.

Focused on the enterprise of science education, the series explores the design and use of technology-enhanced learning environments, applications of artificial intelligence, simulation, and computation, and innovations in online and blended learning, including MOOCs and virtual laboratories. It also considers augmented reality, authentic digital experiences, strategies for supporting diversity and inclusion, research methodologies, workforce development, and the creation and critique of educational policy and leadership.Book proposals for this series may be submitted to the Publishing Editor: Claudia Acuna E-mail: Claudia.Acuna@Springer.com

Menşure Alkış Küçükaydın
Editor

Preparing K-12 Pre-Service Science Teachers for the Age of Artificial Intelligence

Strategies, Innovations, and Classroom Practices

Springer

Editor
Menşure Alkış Küçükaydın
Eregli Faculty of Education
Necmettin Erbakan University
Konya, Türkiye

ISSN 3091-339X ISSN 3091-3403 (electronic)
Advances in Technology-Rich Science Education
ISBN 978-3-032-18029-2 ISBN 978-3-032-18030-8 (eBook)
https://doi.org/10.1007/978-3-032-18030-8

© The Editor(s) (if applicable) and The Author(s), under exclusive license to Springer Nature Switzerland AG 2026, corrected publication 2026

This work is subject to copyright. All rights are solely and exclusively licensed by the Publisher, whether the whole or part of the material is concerned, specifically the rights of translation, reprinting, reuse of illustrations, recitation, broadcasting, reproduction on microfilms or in any other physical way, and transmission or information storage and retrieval, electronic adaptation, computer software, or by similar or dissimilar methodology now known or hereafter developed.
The use of general descriptive names, registered names, trademarks, service marks, etc. in this publication does not imply, even in the absence of a specific statement, that such names are exempt from the relevant protective laws and regulations and therefore free for general use.
The publisher, the authors and the editors are safe to assume that the advice and information in this book are believed to be true and accurate at the date of publication. Neither the publisher nor the authors or the editors give a warranty, expressed or implied, with respect to the material contained herein or for any errors or omissions that may have been made. The publisher remains neutral with regard to jurisdictional claims in published maps and institutional affiliations.

This Springer imprint is published by the registered company Springer Nature Switzerland AG
The registered company address is: Gewerbestrasse 11, 6330 Cham, Switzerland

If disposing of this product, please recycle the paper.

Preface

We live in an age shaped by uncertainty and accelerated by technological transformation. Today's educators are responsible for not only imparting knowledge but also for equipping students with the skills to process, question, and use knowledge responsibly. This book was born precisely at the intersection of this responsibility. It aims to train pre-service teachers not only to use artificial intelligence (AI) but also to understand its social, pedagogical, and ethical dimensions.

AI technologies are no longer abstract predictions of the future; they have become an integral part of daily life, decision-making processes, and, in particular, education. This book argues that science education must be restructured in the age of AI, aiming to equip pre-service teachers not only with instrumental knowledge but also with critical thinking, conceptual awareness, and pedagogical flexibility.

This book incorporates interdisciplinary strategies, classroom-based applications, student-centered models, and examples of technology-supported teaching. Each chapter of the book provides concrete examples of how science education can be reimagined with the aid of AI while also emphasizing fundamental concepts such as scientific literacy, accessibility, and ethical awareness. In the age of social media, while access to information has become easier, misinformation and misconceptions about AI have also become more widespread. In this context, the book aims to equip pre-service teachers not only with technical skills but also with the information literacy and scientific inquiry skills necessary to combat misinformation.

This journey should be viewed as an effort focused on transforming learning rather than the static transfer of knowledge because the future shaped by AI will be a reflection not only of algorithms but also of human intellect, values, and intuition. The ideas presented in this book contribute to positioning pre-service teachers as active, conscious, and equipped individuals in the new era of education.

Overview of the Book

Chapter 1, In the introduction, the book highlights how AI technologies in the second quarter of the 21st century are reshaping education by transforming teacher roles, learning environments, and pedagogical approaches. Focusing on science education, it brings together theoretical and practical studies on AI integration, aiming to prepare pre-service K–12 teachers for digital transformation, foster AI literacy, and guide them in creating ethical, inclusive, and inquiry-based learning environments.

Chapter 2, This chapter examines the integration of artificial intelligence into K–12 science education from multiple perspectives. It discusses AI-TPCK, teacher literacy, ethical and pedagogical issues in AI-supported learning, data security, algorithmic bias, inclusive teaching, AI's role in personalized feedback, student behavior analysis, and learning data interpretation. The chapter also covers AI in STEM education, problem-based and engineering-design learning, ethical leadership in AI, and the future of teacher education.

Chapter 3, This chapter addresses the role of artificial intelligence in science education at a fundamental level, emphasizing the preparation of K-12 science teachers for digital transformation. The integration of AI in education is evaluated in terms of its historical development, theoretical approaches (TAM, UDL, TPACK), and ethical dimensions. Examples are provided using the SAMR model to explain how AI tools can support inclusive, accessible science education that fosters critical thinking. The goal is to ensure that technological advances increase opportunities rather than exacerbate educational inequalities.

Chapter 4, This chapter examines the impact of digital transformation on science education through the integration of artificial intelligence. The mixed-methods research conducted with pre-service teachers and students in Indonesia revealed that pre-service teachers have high ethical awareness, moderate AI literacy, but limited pedagogical application skills. AI-based simulations significantly enhance students' problem-solving and critical thinking skills compared to traditional methods. The study proposes a modular blended learning approach to improve teacher competencies.

Chapter 5, This chapter emphasizes that science education in the age of artificial intelligence aims not only to transfer knowledge but also to develop creativity, ethical thinking, and adaptable problem-solving skills. To support these goals, the Design Thinking (DT) approach is recommended for pre-service teachers to generate solutions to complex classroom problems. In particular, this process—which consists of empathy, problem definition, idea generation, prototyping, and testing—guides the development of student-centered solutions that balance scientific accuracy with social and ethical responsibility. Furthermore, the practical effects of DT are demonstrated through example projects, such as those related to laboratory safety. Finally, future research areas in teacher development and classroom innovation are also presented.

Chapter 6, This chapter presents a practical framework for helping K–12 pre-service teachers use generative AI (GenAI) to design and implement inquiry-based STEM activities. It shows how GenAI can support student exploration, reflection, and collaboration. Instead of focusing on technical training, the chapter emphasizes pedagogical strategies—such as creating open-ended prompts, co-developing rubrics, and guiding student-led investigations. A classroom example illustrates how GenAI can foster autonomy and critical thinking, offering future educators a replicable model for integrating AI into student-centered STEM learning.

Chapter 7, This chapter presents a systematic review of empirical studies (2010–2025) on how AI supports the development of inquiry-based teaching competencies in pre-service science teachers. It identifies key AI tools—such as intelligent tutoring systems, generative chatbots, virtual simulations, and analytics platforms—and examines their impact on skills like questioning, hypothesis generation, data interpretation, and reflective reasoning. While AI shows promise in enhancing these competencies, challenges including limited AI literacy, infrastructural gaps, pedagogical misalignment, and ethical concerns hinder effective integration. The chapter advocates for theoretically grounded, context-sensitive approaches that position AI as a cognitive partner and offers directions for future research.

Chapter 8, This chapter examines how pre-service teachers experience artificial intelligence tools in the instructional design process and how these experiences reflect on their pedagogical approaches, ethical awareness, and perceptions of professional development. Conducted using interpretive phenomenology, a qualitative method, the research involved an AI teaching activity for the primary school level, and data were collected through semi-structured questionnaires. The findings reveal pre-service teachers' levels of awareness and readiness regarding AI applications, teacher roles, pedagogical integration, and ethical and safety issues.

Chapter 9, Synthesizing insights from 18 studies conducted between 2020 and 2025, this chapter critically examines the role of Generative Artificial Intelligence (GenAI) in fostering higher-order thinking skills within K–12 science and mathematics education. When used with structured pedagogies, GenAI enhances problem solving and computational thinking; however, its impact depends on teacher mediation and AI literacy. Sensitive to digital inequalities, GenAI may be less effective than traditional methods in low-resource settings. Consequently, the teacher's role is shifting from content transmitter to epistemic curator, requiring advanced pedagogical, ethical, and technical competencies.

Chapter 10, This chapter introduces the CILP Framework for integrating generative artificial intelligence into STEM education in an ethical and effective manner. This framework positions large language models (e.g., ChatGPT) as cognitive partners that support conceptual reasoning, reflection, and metacognitive development. A quasi-experimental study demonstrated that AI-supported inquiry guided by CILP enhanced students' understanding of thermodynamics and reduced misconceptions. Through continuous guidance, reflection, and evaluation, students gained deeper conceptual integration and epistemic competence. The chapter aims to cultivate AI-ready teachers by translating these findings into AI literacy, guidance design, human—AI hybrid teaching, and ethical awareness strategies for pre-service teachers.

Chapter 11, AI holds significant promise for transforming science education by enabling personalized learning, real-time feedback, data-driven assessment, and experiential learning, while also reshaping the teacher's role. This chapter explores both the long-term potential and the challenges of AI integration, including ethical concerns, privacy issues, teacher preparedness, and equity in access. The insights provided aim to support educators, administrators, and policymakers in leveraging AI to improve science teaching and learning.

Chapter 12, This chapter examines how teachers integrate artificial intelligence into mathematics, science, and computer science courses using three case studies from Korea and the United States. The chapter examines how AI serves as both a supportive and a hindering tool, influencing student cognition and shaping epistemic emotions such as curiosity, confusion, and frustration in inquiry-based learning. The findings show that effective teacher use of AI enhances conceptual understanding, problem-solving, and emotional engagement. Conversely, teacher misuse of AI creates cognitive load and fosters dependency. Cross-cultural comparisons reveal that Korean teachers use structured, concept-focused strategies, while American teachers favor open-ended, collaborative guidance. By emphasizing the mediating role of teachers in human–AI interaction, the study highlights how pedagogical design can support student autonomy, emotional regulation, and critical AI literacy.

Chapter 13, This chapter examines GenAI's transformative potential in science education within the framework of new materialism theory. GenAI is positioned not merely as a tool. It acts as a dialogic partner that supports inquiry, models complexity, and encourages epistemic reflection. GenAI challenges traditional concepts such as knowledge, assessment, and authorship. It requires critical AI literacy encompassing ethical awareness, material interaction, and inclusive representation. The section calls for robust infrastructure, clear policy guidance, and effective professional development for human-centered and critical science education.

Target Audience and Potential Uses

This book is aimed at a broad readership interested in exploring the transformative power of AI in education from pedagogical, ethical, and practical perspectives. This work, which serves as a guide for pre-service science teachers, teaching staff, and academics in education faculties, is also a valuable reference source for many professionals, from experts who design teaching materials to STEM and science education researchers, from decision-makers who shape education policies to teachers who are sensitive to the requirements of the digital age. The book is not only suitable as a core textbook in teacher education programs but also as a valuable resource for educators. Still, it can also be effectively utilized in professional development training, educational technology projects, and the preparation of policy documents. This study, which evaluates AI-supported teaching from a holistic perspective, serves as a contemporary guide for all stakeholders seeking to contribute to scientific thinking, ethical awareness, and inclusivity in education.

Acknowledgements

I would like to express my sincere gratitude to the many individuals who contributed to the realization of this study, entitled *Preparing K–12 Pre-Service Science Teachers for the Age of Artificial Intelligence: Strategies,Innovations,and Classroom Applications*. The combination of scientific and pedagogical contributions, along with the strong support of a publishing house, made the process of creating this book both meaningful and effective.

In particular, the professional support and visionary approach provided by Springer Publishing played a decisive role in bringing this book to the international academic community. I would like to extend my special thanks to Claudia Acuna, whose patience, openness, and constructive feedback throughout the publication process greatly facilitated the process and made our collaboration productive. Our communication not only strengthened the preparation process for publication but also reinforced my enthusiasm for this work to be effective on a global scale.

I also sincerely thank all academic partners, pre-service teachers, and colleagues who embrace innovative approaches in education for their contributions to the creation of this book. Without their contributions and inspiration, this work would not have achieved such a comprehensive and transformative structure.

Menşure Alkış Küçükaydın

Contents

Chapter 1
Introduction

Menşure Alkış Küçükaydın

This introductory section outlines the book's general framework by pursuing the following four main objectives:

- To explain the historical development and theoretical foundations of AI in science education
- To define AI literacy, ethical awareness, and pedagogical integration processes in the context of teacher education
- To analyze the original models presented in the book (AI-TPACK, CILP, DT, etc.) in a comparative manner
- To discuss the effects of AI on learning through cross-cultural application examples and findings based on pre-service teachers' experiences.

Within this framework, both theoretical and empirical contributions will be synthesized throughout the chapter, providing the reader with a comprehensive view of how science education can be restructured in the age of AI. Digital transformation in education systems refers not only to the use of technological tools but also to fundamental changes in the ways information is accessed, processed, and structured. Science education is one of the areas most affected by this transformation. This is because processes such as scientific thinking, data analysis, modeling, and problem solving are directly related to AI technologies. With the introduction of generative AI systems such as ChatGPT, Gemini, and Claude into classroom environments, teachers' role as knowledge transmitters is giving way to that of learning experience designers.

The integration of AI into science education is not only a technical innovation but also an epistemological transformation. The nature of knowledge, the structure of learning processes, and teacher-student interaction are being redefined. In this

M. A. Küçükaydın (✉)
Necmettin Erbakan University, Konya, Turkey
e-mail: mensurealkis@hotmail.com

© The Author(s), under exclusive license to Springer Nature Switzerland AG 2026

M. Alkış Küçükaydın (ed.), *Preparing K-12 Pre-Service Science Teachers for the Age of Artificial Intelligence*, Advances in Technology-Rich Science Education 4,
https://doi.org/10.1007/978-3-032-18030-8_1

context, pre-service teachers must not only be familiar with digital tools but also understand their pedagogical value, ethical boundaries, and inclusive potential.

The chapters in the book address the role of AI in science education from different angles:

- The historical development process (from Turing to ELIZA to ChatGPT)
- Technology adoption models (TAM), universal design principles (UDL), and technological pedagogical content knowledge (TPACK)
- Pre-service teachers' experiences with AI tools, their ethical awareness, and perceptions of professional development
- The impact of AI in the context of higher-order thinking skills (HOTS), epistemic emotions, and inquiry-based learning.

It is emphasized that in this transformation process, teachers should be trained not only as technology users but also as ethical guides, inquiry-based learning designers, and advocates for digital equity.

1.1 Theoretical Foundations: Constructivism, Inquiry-Based Learning, and AI-Based Pedagogical Models

1.1.1 Constructivist Learning and Inquiry-Based Pedagogy

Science education is closely linked to the constructivist learning theory. According to this theory, learning occurs through the active construction of knowledge by the individual (Piaget, 1976; Vygotsky, 1978). Inquiry-based learning is the classroom application of the constructivist approach. Students generate scientific questions, collect data, develop hypotheses, and formulate explanations (Bybee, 2004). The 5E and 7E models, in particular, provide a fundamental framework for science pre-service teachers to design inquiry-based lessons.

The integration of AI into this pedagogical approach can create a learning environment that supports students' scientific reasoning processes, reduces conceptual errors, and increases their metacognitive awareness (Holmes et al., 2019). However, for this integration to be effective, teachers must possess both technological and pedagogical knowledge.

1.1.2 The AI-TPACK Model: Extending Technology, Pedagogy, and Content Knowledge with AI

The TPACK model (Technological Pedagogical Content Knowledge) is widely used to describe teachers' knowledge domains related to technology integration

(Koehler & Mishra, 2009). The AI-TPACK model, on the other hand, reconstructs this framework in the age of AI, defining the types of knowledge specific to AI (AI-TK) and how this knowledge can be integrated with pedagogical (AI-TPK) and content (AI-TCK) contexts (Ning et al., 2024).

Structural equation modeling by Karataş and Ataç (2025) shows that pre-service teachers' AI-TPACK levels are closely related to ethical awareness and pedagogical flexibility. This model enables teachers to view AI tools not merely as technical tools but as cognitive partners that guide their pedagogical decision-making processes.

1.1.3 CILP Framework: Constructivist Inquiry-Based Learning Prompting

The eighth chapter of the book will discuss the Constructivist Inquiry-Based Learning Prompting (CILP) Framework, a unique model aimed at the ethical and effective integration of AI into inquiry-based learning processes. This framework structures the learning process into five stages: prompting, dialogue with AI, reflection, conceptual integration, and evaluation. Each stage aims to increase the learner's epistemic agency and conceptual depth.

The CILP model frames pre-service teachers' interactions with AI not merely as information retrieval, but as processes of inquiry, critical evaluation, and conceptual synthesis. Experimental studies on thermodynamics topics demonstrate that this model is effective in reducing conceptual misunderstandings and enhancing learning coherence.

1.1.4 Higher-Order Thinking Skills (HOTS), AI Literacy, and Conceptual Transformation

The integration of AI technologies into science and mathematics education not only diversifies teaching tools but also has the potential to develop students' higher-order thinking skills (HOTS). These skills include analysis, evaluation, creation, problem-solving, computational thinking, and metacognitive awareness (Facione, 1990; Paul & Elder, 2018). The chapters in the book examine AI's supportive role in HOTS development through both direct measurements and indirect observations.

The impact of AI on HOTS varies depending on the quality of pedagogical design and teacher mediation. The fundamental mechanisms that support HOTS development are not only students obtaining information from AI but also engaging in dialogue with AI, questioning the responses provided, analyzing errors, and generating alternative solutions. Discussing erroneous AI outputs, in particular, triggers students' critical thinking and metacognitive skills (Kasneci et al., 2023; Lin & Yu, 2025).

AI's contribution to HOTS development can be explained through three fundamental mechanisms:

- Socratic Dialogue and Error Analysis: Students' ability to identify, correct, and justify errors in AI outputs strengthens critical thinking and epistemic agency (UNESCO, 2025).
- Adaptive Support and Feedback Loops: AI-powered guidance tailored to students' individual needs develops problem-solving and computational thinking skills (Grover & Pea, 2013).
- Creative Production and Conceptual Modeling: Students generating new ideas, designing alternative scenarios, and creating scientific models with AI encourages creative thinking (Gibson, 1992; Lipman, 2003).

However, for this potential to be realized, teachers must position AI not merely as a source of information but as an active partner in the learning process. Teacher guidance is essential for students to question, verify, and contextualize AI output (Tang et al., 2024).

AI literacy is one of the prerequisites for HOTS development. Students' understanding of how AI systems work, what data sources they rely on, what biases they may carry, and what limitations they have forms the basis of critical digital literacy (Cooper & Tang, 2024; Ng et al., 2024). In this context, pre-service teachers must view AI literacy not only as a technical skill but also as an ethical, epistemological, and pedagogical area of competence.

The book's chapters demonstrate AI's contribution to HOTS development across different contexts. In some classes, students gained conceptual coherence by solving scientific problems with AI; in some cases, AI's limited accuracy prompted students to exercise their inquiry skills. This diversity shows that AI can be positioned as a tool that not only provides information in learning environments but also provokes thinking.

1.2 Accessibility, Inclusivity, and Ethical Principles: AI Integration within the UDL and SAMR Frameworks

The integration of AI technologies into science education should be evaluated not only for pedagogical efficiency but also for accessibility, inclusivity, and ethical responsibility. The book's chapters emphasize the importance of integrating AI in a way that is sensitive to diverse learning styles, special needs, and cultural contexts. In this context, the Universal Design for Learning (UDL) and Substitution–Augmentation–Modification–Redefinition (SAMR) models provide functional frameworks for evaluating both the depth and inclusivity of AI integration.

1.2.1 UDL Principles and AIIntegration

UDL is a neuroscience-based framework that aims to make learning environments accessible to all students (Roski et al., 2024). Its three core principles—engagement, representation, and action & expression—directly align with AI technologies.

- Engagement: AI tools can increase motivation by offering content that can be customized according to students' interests and cultural contexts. For example, in a class addressing climate change, students can analyze environmental data specific to their regions using AI-powered simulations.
- Representation: AI systems can cater to different learning styles by presenting scientific concepts in visual, auditory, and tactile forms (Hyatt & Owenz, 2024). In a physics class, wave motion can be presented through graphics, animations, and audio explanations.
- Expression: Students can express what they have learned through different means such as written narratives, AI-supported video presentations, digital posters, or 3D modeling. This diversity enriches learning outcomes and ensures inclusive assessment processes.

Integrating AI with UDL principles makes the learning process more accessible and meaningful, especially for students with special needs. The examples in the book show how AI makes applications such as audio description for visually impaired students and eye-tracking control for those with motor skill difficulties possible.

1.2.2 Depth of AI Integration Using the SAMR Model

The SAMR model defines the level of technology integration into teaching processes in four stages: Substitution, Augmentation, Modification, and Redefinition (Puentedura, 2006). The book's application examples demonstrate how AI can be used at all levels of the SAMR model. For example, if a chemistry teacher uses AI-supported molecular modeling software solely for visualization purposes, this is at the "augmentation" level. However, if students use these models to develop new hypotheses and discuss them with AI, this indicates the 'modification' or "redefinition" level.

1.2.3 Ethical Principles and Responsibility in AI

The use of AI in education also entails ethical responsibilities, including data privacy, algorithmic bias, academic integrity, and equal access. Teachers need to provide guidance on issues such as protecting students' personal data, questioning AI system decisions, and citing sources in content production (Baker & Hawn, 2022; Reidenberg & Schaub, 2018). Algorithmic biases, in particular, can lead to the systematic exclusion of disadvantaged groups. Therefore, it is important for pre-service teachers to have the ethical reflexes to question how AI systems are trained, what data sets they rely on, and which groups they do not represent. The book's chapters emphasize that pre-service teachers should consider not only utilitarian approaches but also social justice, inclusivity, and human-centered design in their ethical decision-making. In this context, it is recommended that teacher education programs include components such as ethical awareness, digital citizenship, and data literacy.

1.2.4 Teacher Education and AI Literacy: From Conceptual Competence to Pedagogical Practice

The integration of AI technologies into science education requires teachers to possess not only technical knowledge but also pedagogical, ethical, and epistemological competencies. The chapters presented in the book highlight four key competency areas that guide pre-service teachers 'interactions with AI: (1) prompt engineering, (2) AI literacy and interpretation skills, (3) hybrid teaching design, and (4) digital ethics and inclusivity awareness.

1.2.5 Conceptual Competence: AI Literacy and System Knowledge

AI literacy requires teachers to understand how AI systems work, the data sources they rely on, how they learn, and their limitations (Ng et al., 2024). This knowledge enables teachers to evaluate the accuracy of AI outputs, recognize biases, and guide students toward critical digital citizenship. It is important for pre-service teachers, especially when interacting with large language models (LLMs), to understand the system's probabilistic response-generation logic and to recognize inconsistencies, omissions, or "hallucinations" in responses (Walter, 2024). This awareness enables AI to be used not only as a source of information but also as a tool that triggers questioning and thinking processes.

1.2.6 Ethical Competence: Digital Responsibility and Inclusivity

The integration of AI into educational settings requires teachers to be aware of ethical issues such as data privacy, algorithmic bias, academic integrity, and equal access (Baker & Hawn, 2022; Reidenberg & Schaub, 2018). Pre-service teachers should provide guidance on topics such as protecting students' personal data, questioning AI system decisions, and citing sources in content production.

1.2.7 Application Examples: Microteaching, Reflective Portfolios, and Scenario Analyses

The book's application examples enable pre-service teachers to experience their interactions with AI in a structured manner. Microteaching applications allow pre-service teachers to test AI-supported lesson scenarios with small groups and receive feedback. Reflective portfolios allow candidates to document their interactions with AI, their ethical assessments, and their pedagogical decisions. Scenario analyses encourage critical thinking by opening up discussion about the effects of AI in the classroom. These applications support pre-service teachers in developing not only technical skills but also pedagogical reasoning, ethical reflection, and conceptual integrity.

References

Baker, R. S., & Hawn, A. (2022). Algorithmic bias in education. *International Journal of Artificial Intelligence in Education, 32*(4), 1052–1092. https://doi.org/10.1007/s40593-021-00285-9

Bybee, R. W. (2004). Scientific inquiry and science teaching. In L. B. Flick & N. G. Lederman (Eds.), *Scientific inquiry and the nature of science* (pp. 1–14). Kluwer.

Cooper, G., & Tang, K. S. (2024). Pixels and pedagogy: Examining science education imagery by generative artificial intelligence. *Journal of Science Education and Technology, 33*(4), 556–568. https://doi.org/10.1007/s10956-024-10104-0

Facione, P. A. (1990). *Critical thinking: A statement of expert consensus for purposes of educational assessment and instruction.* The California Academic Press.

Gibson, S. (1992). Serious creativity: Using the power of lateral thinking to create new ideas. *Industry Week, 241*(16), 43–44.

Grover, S., & Pea, R. (2013). Computational thinking in K–12: A review of the state of the field. *Educational Researcher, 42*(1), 38–43. https://doi.org/10.3102/0013189X12463051

Holmes, W., Bialik, M., & Fadel, C. (2019). *Artificial intelligence in education: Promises and implications for teaching and learning.* Center for Curriculum Redesign.

Hyatt, S. E., & Owenz, M. B. (2024). Using universal design for learning and artificial intelligence to support students with disabilities. *College Teaching*, 1-8. https://doi.org/10.1080/87567555.2024.2313468

Karataş, F., & Ataç, B. A. (2025). When TPACK meets artificial intelligence: Analyzing TPACK and AI-TPACK components through structural equation modelling. *Education and Information Technologies, 30*(7), 8979–9004. https://doi.org/10.1007/s10639-024-13164-2

Kasneci, E., Seßler, K., Küchemann, S., Bannert, M., Dementieva, D., Fischer, F., & Kasneci, G. (2023). ChatGPT for good? On opportunities and challenges of large language models for education. *Learning and Individual Differences, 103*, 102274. https://doi.org/10.1016/j.lindif.2023.102274

Koehler, M., & Mishra, P. (2009). What is technological pedagogical content knowledge (TPACK)? *Contemporary Issues in Technology and Teacher Education, 9*(1), 60–70.

Lin, Y., & Yu, Z. (2025). Learner perceptions of artificial intelligence-generated pedagogical agents in language learning videos: Embodiment effects on technology acceptance. *International Journal of Human-Computer Interaction, 41*(2), 1606–1627. https://doi.org/10.1080/10447318.2024.2359222

Lipman, M. (2003). *Thinking in education.* Cambridge University Press.

Ng, D. T. K., Su, J., Leung, J. K. L., & Chu, S. K. W. (2024). Artificial intelligence (AI) literacy education in secondary schools: A review. *Interactive Learning Environments, 32*(10), 6204–6224. https://doi.org/10.1080/10494820.2023.2255228

Ning, Y., Zhang, C., Xu, B., Zhou, Y., & Wijaya, T. T. (2024). Teachers' AI-TPACK: Exploring the relationship between knowledge elements. *Sustainability, 16*(3), 978. https://doi.org/10.3390/su16030978

Paul, R., & Elder, L. (2018). *The miniature guide to critical thinking: Concepts and tools* (8th ed.). Foundation for Critical Thinking.

Piaget, J. (1976). *The grasp of consciousness: Action and concept in the young child.* Harvard University Press.

Puentedura, R. (2006, August). Transformation, technology, and education. http://hippasus.com/resources/tte/

Reidenberg, J. R., & Schaub, F. (2018). Achieving big data privacy in education. *Theory and Research in Education, 16*(3), 263–279. https://doi.org/10.1177/1477878518805308

Roski, M., Sebastian, R., Ewerth, R., Hoppe, A., & Nehring, A. (2024). Learning analytics and the Universal Design for Learning (UDL): A clustering approach. *Computers & Education, 214*, Article 105028. https://doi.org/10.1016/j.compedu.2024.105028

Tang, A., Li, K. K., Kwok, K. O., Cao, L., Luong, S., & Tam, W. (2024). The importance of transparency: Declaring the use of generative artificial intelligence (AI) in academic writing. *Journal of Nursing Scholarship, 56*(2), 314–318.

UNESCO. (2025). *Artificial intelligence and education: Guidance for policy makers.* United Nations Educational, Scientific and Cultural Organization.

Vygotsky, L. S. (1978). *Mind in society: The development of higher psychological processes* (M. Cole, V. John-Steiner, S. Scribner, & E. Souberman, Eds. & Trans.). Harvard University Press.

Walter, Y. (2024). Embracing the future of Artificial Intelligence in the classroom: The relevance of AI literacy, prompt engineering, and critical thinking in modern education. *International Journal of Educational Technology in Higher Education, 21*(1), 15. https://doi.org/10.1186/s41239-024-00448-3

Chapter 2
Foundations of Artificial Intelligence in Science Education

Menşure Alkış Küçükaydın

2.1 Introduction

The first quarter of the twenty-first century has been a period during which artificial intelligence (AI)-based technologies have rapidly expanded in the field of education, fundamentally transforming teaching processes. This transformation necessitates a redefinition of teacher competencies, particularly in knowledge-based and skill-oriented disciplines, such as science education (Luckin et al., 2016). K-12 science pre-service teachers should be trained not only as individuals familiar with technological tools but also as professionals who can effectively integrate these technologies for pedagogical purposes, develop ethical awareness, and think critically.

The integration of AI into educational environments reveals that digital transformation is not only a technical change but also an epistemological one. Education systems are reimagining the nature of accessing, processing, and designing learning processes (Selwyn, 2019). In this context, pre-service teachers must not only be proficient in the use of digital tools but also possess conceptual and theoretical knowledge about the pedagogical value of these tools. The types of knowledge that teachers need to possess in the age of artificial intelligence can be explained by the Technological Pedagogical Content Knowledge (TPACK) framework. The TPACK model envisions teachers establishing a holistic balance between content knowledge, pedagogical knowledge, and technological knowledge (Koehler & Mishra, 2009). However, the introduction of AI-based teaching tools has made this model even more dynamic. Current literature emphasizes that teachers' acquisition of "AI literacy" is a fourth dimension that needs to be integrated into the TPACK model (Umar et al., 2025).

M. Alkış Küçükaydın (✉)
Necmettin Erbakan University, Konya, Turkey
e-mail: mensurealkis@hotmail.com

© The Author(s), under exclusive license to Springer Nature Switzerland AG 2026
M. Alkış Küçükaydın (ed.), *Preparing K-12 Pre-Service Science Teachers for the Age of Artificial Intelligence*, Advances in Technology-Rich Science Education 4,
https://doi.org/10.1007/978-3-032-18030-8_2

AI literacy encompasses not only the use of technology but also its subcomponents, including algorithmic thinking, data-driven decision-making, ethical evaluation, and human-centered technology design (Touretzky et al., 2019). Therefore, programs developed for science pre-service teachers should integrate interdisciplinary knowledge areas and equip them not only with technical skills but also with the ability to critically examine the social implications of AI systems. AI-supported learning environments have the potential to deepen students'conceptual understanding processes. For example, adaptive learning systems that work with AI can analyze students' cognitive profiles and provide them with personalized content (Holmes et al., 2019). Similarly, interactive digital assistants powered by natural language processing (NLP) technologies support inquiry-based learning by answering students' questions about scientific concepts (Chen et al., 2020). Simulations, virtual laboratories, and augmented reality applications commonly used in science education also offer more intuitive, interactive, and data-driven learning experiences when developed with AI algorithms. For example, AI-based systems that change the course of a simulation based on the choices made by the student during the experiment support conceptual transfer and encourage scientific reasoning (Zhai et al., 2021).

With this transformation, teachers' traditional role as "knowledge transferors" is being replaced by that of "learning experience designers" and "digital guides" (Ertmer & Ottenbreit-Leftwich, 2013). AI systems can analyze student performance, deliver content, and provide feedback; however, evaluating the pedagogical accuracy of these processes and monitoring their suitability for students remains the teacher's responsibility. In this context, it is necessary to train pre-service teachers to be individuals who approach technology critically and who are mindful of social equality and ethical principles. The unquestioning adoption of educational technologies can weaken pedagogical depth and increase inequality of opportunity among students (Williamson & Eynon, 2020). Therefore, pre-service teachers need to be reflective practitioners who continually evaluate the impact of technology on learning processes.

To prepare science pre-service teachers for digital transformation, multi-layered structures should be incorporated into teacher education programs. These structures can be organized around (1) theoretical knowledge (basic principles of AI, educational technology theory), (2) applied content (classroom scenarios using AI-supported tools), (3) ethical and social dimensions (algorithmic biases, data security) and (4) professional development (continuous learning strategies with AI) (Zawacki-Richter et al., 2019). In addition, microteaching applications should be developed to demonstrate to pre-service teachers how to utilize these tools not only in class but also in processes ranging from lesson planning to assessment. In these processes, the scenario-based use of AI-supported learning management systems (LMS), automatic assessment platforms, and analytical reports will be highly functional.

2.2 Science Education in the Age of Artificial Intelligence: Preparing K–12 Pre-Service Teachers for Digital Transformation

One of the key components in preparing pre-service teachers for digital transformation is technology literacy. However, today, this concept seems to be increasingly giving way to "AI literacy." AI literacy requires not only an understanding of how technological tools work but also a comprehension of their decision-making processes, algorithmic structures, data-driven operations, and ethical implications (Touretzky et al., 2019). It is particularly important for science teachers working at the K–12 level to understand how AI systems function, how they process student data, and the impact of these processes on learning outcomes. In this context, pre-service teachers must possess both technical and critical thinking skills to make pedagogical decisions in AI-supported learning environments (Zawacki-Richter et al., 2019).

The TPACK model, widely used to explain teachers' knowledge areas related to technology integration, is being reinterpreted in the era of AI. Developed by Koehler and Mishra (2009), this model envisions teachers establishing a holistic balance between content knowledge (CK), pedagogical knowledge (PK), and technological knowledge (TK). However, the integration of AI into educational environments has necessitated an expansion of this model. The AI-TPACK model, developed by Ning et al. (2024), reconfigures the TPACK components to explain how teachers can use AI technologies in a pedagogical context. According to this model, teachers must possess not only technological knowledge but also AI-specific knowledge types (AI-TK) and knowledge about how this knowledge can be integrated into pedagogical (AI-TPK) and content-related (AI-TCK) contexts. Structural equation modeling has shown that pre-service teachers' AI-TPACK levels are closely related to ethical awareness and pedagogical flexibility (Karataş & Ataç, 2024).

2.2.1 Classroom Scenarios: Applications for AI Integration

The integration of AI into science classes enables pre-service teachers to concretize abstract concepts and students to develop scientific thinking skills. For example, AI-supported simulation software used in a middle school science class can enable students to observe the process of photosynthesis under different environmental conditions. Such applications support students in developing scientific process skills, such as hypothesis formation, observation, and conclusion drawing (Chen et al., 2020). In another scenario, a pre-service teacher can use AI-supported learning analytics to identify students' conceptual misunderstandings and restructure their teaching strategies accordingly. Such data-driven teaching practices make the teacher's in-class decision-making processes more conscious and effective (Holmes et al., 2019).

Additionally, AI-supported chatbots can facilitate the learning process by providing instant answers to students' scientific questions. However, for such applications to gain pedagogical value, pre-service teachers need to know how to guide these systems, when to activate them, and how to support students'critical thinking skills (Mishra et al., 2023). Therefore, science education in the AI era is not only about bringing technological tools into the classroom but also about contextualizing these tools within pedagogical, ethical, and epistemological frameworks. Preparing K–12 science pre-service teachers for digital transformation requires a teacher education approach that goes beyond technology literacy and encompasses holistic knowledge areas such as AI literacy, TPACK, and AI-TPACK. In this context, the curricula to be developed should prepare pre-service teachers not only for today's learning environments but also for those of the future.

2.3 Ethical and Pedagogical Approaches in AI Artificial Intelligence (AI) -Supported Learning

The integration of AI technologies into educational environments presents significant opportunities for enhancing teaching processes, but it also entails serious ethical and pedagogical responsibilities. In particular, protecting privacy in the collection, processing, and interpretation of student data, ensuring that algorithmic decisions are fair, and inclusively designing learning environments are among the key issues that pre-service teachers need to consider (Nguyen et al., 2023). In this context, ethical dilemmas encountered in AI-supported learning environments, algorithmic biases, principles of inclusivity, and teachers' ethical decision-making processes are addressed in a multidimensional manner.

2.3.1 A. Student Privacy and Data Protection

AI systems require large amounts of student data to function effectively. This data may include sensitive content such as academic performance, behavioral tendencies, social interactions, and even biometric information (Huang, 2023). However, this situation poses serious risks that threaten students' rights to privacy. In particular, "secondary exploitation" (secondary exploitation), which refers to the use of data for other purposes without the student's knowledge, poses a serious ethical problem (Tan & Yang, 2019).

Regulations such as the European Union's General Data Protection Regulation (GDPR) require the principles of explicit consent, data minimization, and transparency in the collection and processing of student data. However, there are serious shortcomings in the implementation of these principles in many educational institutions (Nguyen et al., 2023). In this context, pre-service teachers must be trained to be conscious individuals not only in terms of pedagogy but also in terms of data ethics.

2.3.2 B. Algorithmic Bias and Equity in Education

The algorithms used by AI systems in decision-making processes are often trained with data that reflects the biases of developers or social inequalities. This situation can lead to the systematic exclusion of disadvantaged groups (Baker & Hawn, 2022). For example, some learning analytics systems have been shown to tend to classify students with low socioeconomic backgrounds as "at risk" (Hu & Rangwala, 2020). Such classifications can hinder students' access to support services and undermine educational equity.

Baker and Hawn (2022) note that algorithmic bias in education most frequently manifests in variables such as gender, ethnicity, disability status, and socioeconomic status. In this context, pre-service teachers must understand how algorithms work, comprehend how biases are formed, and develop a critical perspective that enables them to examine and critically question these biases.

2.3.3 C. Inclusive Teaching and the Role of AI

When designed correctly, AI technologies can contribute to the creation of inclusive teaching environments. Applications developed specifically for students with special needs, such as audio description, automatic subtitling, and personalized content delivery, increase participation in the learning process (Melo-López et al., 2025). However, for these technologies to be effective, teachers must integrate the principles of inclusivity into their pedagogical designs.

Mollick and Mollick (2023) emphasize that AI is a valuable tool for developing teaching strategies that can be tailored to different learning styles and individual needs. However, it should not be forgotten that these strategies must be based not only on technical but also on pedagogical foundations. Pre-service teachers should be encouraged to view AI not merely as an "auxiliary tool" but as a "pedagogical partner" capable of transforming the learning process.

2.3.4 D. Ethical Decision-Making Models and the Role of Teachers

In AI-supported learning environments, teachers are not only content providers but also ethical decision-makers. In this context, pre-service teachers need to be familiar with theoretical models that can guide their ethical decision-making processes. Karakuş et al. (2025) found that teachers generally adopt approaches such as virtue ethics, deontological ethics, and social justice ethics when faced with ethical dilemmas. In contrast, AI systems tend to make more utilitarian and outcome-oriented decisions.

This difference suggests that ethical conflicts arise in teachers' interactions with AI systems. For example, while an AI system may automatically classify a student as "at risk" due to poor performance, a teacher may make a different assessment by considering the student's social context and background. In such cases, pre-service teachers need to have ethical reflexes that enable them to question the decisions of AI systems (Nguyen et al., 2023). For AI-supported learning environments to be ethically and pedagogically sustainable, pre-service teachers must approach these technologies not only as users but also as critical evaluators and ethical guides. Protecting privacy, recognizing algorithmic biases, developing inclusive teaching strategies, and imparting ethical decision-making skills have become indispensable components of contemporary teacher education. In this context, it is of great importance to train pre-service teachers to be ethical leaders who will shape not only today's but also tomorrow's educational environments.

2.4 AI-Based Student Models and Learning Analytics

AI-supported educational technologies are being increasingly used to understand student learning processes better and optimize teaching strategies. In particular, student tracking systems, customized feedback mechanisms, behavioral analysis techniques, and learning data interpretation processes are among the most important innovations offered by AI in education (Baker & Hawn, 2022). Therefore, how AI-based student models work, the role of learning analytics in education, and how teachers can use these technologies pedagogically should also be addressed.

AI-supported student monitoring systems analyze students' learning processes in real time, providing valuable information to teachers and educators. These systems can track students' interactions with course materials, assignment completion times, exam performance, and even classroom behavior (Hu & Rangwala, 2020; Hu et al., 2024). AI algorithms integrated with LMS, in particular, identify students' learning tendencies, providing teachers with the opportunity to develop personalized teaching strategies. Student monitoring systems can predict students' future academic success by utilizing predictive analytics techniques and implementing early intervention mechanisms. For example, the AI-supported academic advising system implemented at Georgia State University analyzes students' course performance to identify at-risk students and provides teachers with recommendations for early intervention (Baker & Hawn, 2022). Such systems enable teachers to provide proactive support to students. However, the widespread use of student monitoring systems also raises significant ethical concerns regarding privacy and data ethics. In particular, the long-term tracking of students' learning habits raises serious concerns about data security (Huang, 2023). Teachers must act by the principle of privacy when using such systems and ensure that students' data is consciously protected.

AI-powered feedback systems enhance the learning process by providing personalized feedback tailored to each student's individual learning needs. While traditional feedback mechanisms are often general and delayed, AI-based systems provide instant and personalized feedback, enabling students to identify and correct their mistakes more quickly (Mollick & Mollick, 2023). AI systems that utilize NLP technologies can analyze students' written work and identify grammatical errors, conceptual gaps, and argumentative structure issues. A study conducted at Stanford University shows that AI-supported feedback systems reduce teachers' workload while increasing students' academic achievement (Cuéllar et al., 2025). Thus, these systems enable teachers to provide more targeted feedback to students, helping them improve their writing skills. Adaptive learning systems analyze students' learning profiles and offer dynamic teaching methods. AI-based learning platforms can assess students' strengths and weaknesses and recommend customized materials and exercises (Salas-Pilco et al., 2023). Such systems provide students with the opportunity to progress at their own pace and experience personalized education throughout the learning process.

AI-supported behavioral analysis systems analyze students' interactions in the classroom and online learning environments, providing teachers with more comprehensive information. Techniques such as eye tracking, facial expression analysis, and voice tone assessment are used to determine how actively students participate in the learning process (Li et al., 2023). For example, the study titled Real-Time Classroom Behavior Analysis reveals that AI-supported systems analyze students' attention levels in the classroom and provide teachers with instant feedback (Hu et al., 2024). Such systems enable teachers to adapt their teaching techniques according to students' varying attention levels. Learning analytics is an approach that utilizes big data techniques to gain a deeper understanding of students' learning processes and optimize teaching strategies. AI-supported learning analytics analyze students' interactions with course materials, exam performance, and learning speeds, providing teachers with the opportunity to make data-driven decisions (Salas-Pilco et al., 2023). Learning analytics can help teachers identify students' conceptual misunderstandings and adapt their teaching strategies accordingly. For example, the AI-supported learning analytics system implemented at Purdue University analyzes students' access times to course materials and identifies topics that require further explanation for teachers (Microsoft, 2021).

In this context, AI-based student models and learning analytics emerge as crucial tools that enhance data-driven decision-making processes in education. Student monitoring systems, customized feedback mechanisms, behavioral analysis techniques, and learning data interpretation processes help teachers better understand students' learning processes and optimize their teaching strategies. However, for these technologies to be ethically and pedagogically sustainable, pre-service teachers must critically evaluate AI systems and develop ethical awareness.

2.5 The Impact of AI on STEM Integration

AI technologies are fundamentally transforming the nature of STEM (Science, Technology, Engineering, and Mathematics) education, offering new pedagogical approaches that support interdisciplinary learning processes. The integration of AI into STEM education is evaluated through key components, including interdisciplinary scenarios, problem-based learning (PBL), the engineering design process, and systems thinking (Xu & Ouyang, 2022). In this context, this book examines the impact of AI on STEM education, how it transforms the pedagogical approaches of pre-service teachers, and how it develops students' scientific thinking skills.

STEM education has a structure that encourages interdisciplinary learning processes by transcending traditional course boundaries. The integration of AI into STEM education brings together different disciplines to support complex problem-solving and scientific modeling processes (Cai et al., 2024). For example, bioinformatics applications that combine biology and computer science enable students to analyze genetic data and model biological processes using AI algorithms. AI-supported interdisciplinary scenarios enable students to integrate their knowledge and skills across various fields, including data science, robotics, environmental sciences, and mathematical modeling. In particular, AI-supported simulations enable students to analyze complex systems and evaluate the effects of different variables (Xu & Ouyang, 2022). Such applications strengthen the processes of experimental learning and hypothesis testing in STEM education.

PBL is a pedagogical approach that enables students to learn by solving real-world problems. The integration of AI into PBL processes helps students develop their data analysis, hypothesis development, and solution-generation skills (Ruiz Viruel et al., 2025). For example, students can use AI-supported environmental analysis systems to examine air pollution data and develop sustainable solutions. AI-supported PBL applications enable students to develop critical thinking and collaborative learning skills. In particular, AI-based learning analytics provide teachers with feedback by analyzing students' problem-solving processes. Such systems help students manage their learning processes more consciously and systematically.

The engineering design process is a fundamental approach in STEM education that enables students to acquire skills in creative thinking, prototyping, and system development. The integration of AI into the engineering design process helps students develop their design optimization, automatic modeling, and data-driven decision-making skills (Khanolkar et al., 2023). For example, students can develop engineering projects and optimize their designs using AI-supported CAD

(Computer-Aided Design) systems. AI algorithms assist students in analyzing their design processes and evaluating factors such as material selection, energy efficiency, and production processes (Chien & Yao, 2020). Such systems enable students to make more informed and data-driven decisions in their engineering projects. AI-supported engineering-design processes strengthen students' experimental learning skills. In particular, AI-based simulations allow students to test and improve their designs in a virtual environment (Belghith et al., 2025). Such applications support creative problem-solving and system development processes in STEM education.

Systemic thinking is an approach that enables students to analyze complex systems and understand the relationships between different components. The integration of AI into STEM education helps students develop their skills in system modeling, data analysis, and predictive simulation (Xu & Ouyang, 2022). For example, students can utilize AI-supported ecological modeling systems to examine the impact of environmental variables on ecosystems. Such systems help students understand causal relationships and gain a deeper comprehension of complex systems. AI-supported systems thinking applications enable students to perform multidisciplinary analysis and make data-driven decisions. The integration of AI into systemic thinking processes enables students to develop predictive analysis and scenario modeling skills, enhancing their ability to anticipate and respond to complex challenges. In particular, AI-based decision support systems enable students to analyze various scenarios and identify the most suitable solutions (Xu & Ouyang, 2022). Such systems support strategic thinking and complex problem-solving processes in STEM education.

In this context, the integration of AI into STEM education brings about a significant transformation that supports interdisciplinary learning processes, strengthens problem-based learning, optimizes engineering design processes, and develops systemic thinking skills. Students become more conscious and creative individuals in STEM fields by acquiring skills such as data analysis, hypothesis development, design optimization, and modeling complex systems in AI-supported learning environments. In this context, it is of great importance for pre-service teachers to understand how to use AI technologies pedagogically and to integrate these technologies in an ethical, inclusive, and sustainable manner.

2.6 Ethical Leadership with AI Artificial Intelligence (AI) in Education and the Future of Teacher Education

The increasing use of AI technologies in education necessitates a restructuring of teacher education programs, with the principles of ethical leadership playing a central role in this process. The sustainable integration of AI in education should be underpinned not only by the development of technical competencies but also by the principles of ethical awareness, inclusiveness, and pedagogical sustainability (Khan et al., 2025). In this context, the final part of the chapter discusses how AI-based content should be structured in teacher training programs, how sustainable education policies can be developed, and how AI can be effectively utilized in teachers' professional development processes.

To utilize AI effectively in education, pre-service teachers must develop a pedagogical and ethical understanding of these technologies. In the AI Competency Framework for Teachers published by UNESCO, four main competency areas were identified for teachers to develop AI literacy: (1) understanding the basic principles of AI, (2) implementing AI-supported teaching strategies, (3) evaluating the ethical and social implications of AI, (4) comprehending AI-related policies and regulations (UNESCO, 2024). The integration of AI-based content in teacher education programs should not be limited to the acquisition of technical skills. Pre-service teachers must understand how to use technologies such as AI-supported learning analytics, adaptive learning systems, and automated assessment tools in a pedagogical context (European Commission, 2022). Furthermore, teachers need to be able to assess how AI systems drive decision-making processes and their impact on equal opportunities in education.

The sustainable use of AI in education is possible not only through the development of technological infrastructure but also through the consideration of ethical and environmental factors. In particular, energy consumption and data processing processes of AI-supported education systems should be carefully considered in terms of sustainability. AI algorithms operating on large datasets can lead to high energy consumption and an increase in carbon footprint (EY, 2025). Therefore, teacher training programs should provide training on sustainable AI applications, and teachers should be encouraged to select AI technologies that consider environmental impacts.

For teachers to utilize AI technologies effectively, they must participate in ongoing professional development programs. The ISTE AI Professional Development study suggests the following three main strategies to improve teachers' AI literacy: (1) basic training to understand how to use AI in education, (2) hands-on experience of teachers with AI-supported teaching tools, and (3) professional learning communities to discuss the ethical and pedagogical implications of AI (ISTE, 2023). Teachers' understanding of how to use AI technologies in a pedagogical context should not be limited to acquiring technical skills. According to the report "Professional Development for Teachers in the Age of AI," published by the European Schoolnet Academy, approaches such as PBL, interdisciplinary teaching, and data-driven decision-making should be adopted by teachers to develop AI-supported teaching strategies (Cukurova et al., 2024).

For AI to be used ethically and sustainably in education, policymakers need to develop comprehensive regulations for these technologies. The UNESCO AI and Education Guidance for Policymakers offers the following four main recommendations for the integration of AI into education policies: (1) establishing ethical frameworks for the use of AI in education, (2) preparing national training programs to improve teachers' AI literacy, (3) implementing AI-supported learning analytics in a way that supports equal opportunities in education, (4) using AI in a transparent and accountable manner in education (UNESCO, 2021). Furthermore, in the study

5 Key Policy Ideas to Integrate AI in Education Effectively published by the World Economic Forum, the following five key policy recommendations are presented for the integration of AI into education systems: (1) establishing AI-focused education task forces, (2) integrating AI literacy into the curriculum, (3) providing professional development support for teachers to develop AI-supported teaching strategies, (4) preparing guidelines for the ethical use of AI in education, (5) aligning AI research and development processes with education policies (World Economic Forum, 2024).

Therefore, the integration of AI into ethical leadership in education and teacher education processes should be supported not only by the development of technical skills but also by the principles of sustainability, inclusiveness, and ethical awareness. Addressing AI-based content in teacher education programs in pedagogical and ethical contexts will enable teachers to use these technologies consciously and responsibly. Moreover, structuring educational policies to encourage the ethical use of AI will contribute to teachers' effective adoption of AI-supported learning strategies in their professional development processes.

2.7 Conclusion

The integration of AI technologies in education brings about a broad transformation, ranging from teacher education to ethical leadership and from interdisciplinary STEM practices to learning analytics. AI enables teachers to restructure their pedagogical approaches, making student learning processes more individualized, data-driven, and systematic. However, the sustainable, inclusive, and ethical application of these technologies requires pre-service teachers not only to acquire technical skills but also to develop critical thinking and ethical awareness. Education policies in the AI era must include regulations that support teachers' professional development, data security, and equal opportunity. The integration of teacher training programs with models such as AI-TPACK will position future educators not only as individuals who utilize technology but also as leaders who consciously direct its use. The role of AI in education is no longer just a digital tool; it has become a fundamental structure that guides teachers' decision-making processes, learning experience designs, and the sustainability of education systems. The most critical point in this transformation process is that AI is seen as a tool to support education, but the human factor remains at the center of learning and teaching. Scientific thinking, ethical leadership, and pedagogical innovation are essential for a strong teacher identity in the age of AI.

References

Baker, R. S., & Hawn, A. (2022). Algorithmic bias in education. *International Journal of Artificial Intelligence in Education, 32*(4), 1052–1092. https://doi.org/10.1007/s40593-021-00285-9

Belghith, Y., Riedl, M., Moore, R., Alemdar, M., & Roberts, J. (2025). Exploring AI intervention points in high-school engineering education: A research through co-design approach. *Information and Learning Sciences*. https://doi.org/10.1108/ILS-04-2024-0042
Cai, C., Zhu, G., & Ma, M. (2024). A systemic review of AI for interdisciplinary learning: Application contexts, roles, and influences. *Education and Information Technologies*, 1-47. https://doi.org/10.1007/s10639-024-13193-x
Chen, L., Chen, P., & Lin, Z. (2020). Artificial intelligence in education: A Review. *IEEE Access, 8*, 75264–75278. https://doi.org/10.1109/ACCESS.2020.2988510
Chien, Y. H., & Yao, C. K. (2020). Development of an ai userbot for engineering design education using an intent and flow combined framework. *Applied Sciences, 10*(22), 7970. https://doi.org/10.3390/app10227970
Cuéllar, Ó., Contero, M., & Hincapié, M. (2025). Personalized and timely feedback in online education: Enhancing learning with deep learning and large language models. *Multimodal Technologies and Interaction, 9*(5), 45. https://doi.org/10.3390/mti9050045
Cukurova, M., Kralj, L., Hertz, B., & Saltidou, E. (2024). *Professional development for teachers in the age of AI*. European Schoolnet: Brussels, Belgium.
Ertmer, P. A., & Ottenbreit-Leftwich, A. (2013). Removing obstacles to the pedagogical changes required by Jonassen's vision of authentic technology-enabled learning. *Computers & Education, 64*, 175–182. https://doi.org/10.1016/j.compedu.2012.10.008
European Commission. (2022). *Ethical guidelines on the use of artificial ıntelligence and data in teaching and learning for educators*. https://data.europa.eu/doi/https://doi.org/10.2766/153756
EY. (2025). *AI and sustainability: Opportunities, challenges, and impact*. https://www.ey.com/en_nl/insights/climate-change-sustainability-services/ai-and-sustainability-opportunities-challenges-and-impact
Holmes, W., Bialik, M., & Fadel, C. (2019). *Artificial ıntelligence in education: Promises and ımplications for teaching and learning*. Center for Curriculum Redesign.
Hu, J., Huang, Z., Li, J., Xu, L., & Zou, Y. (2024). Real-time classroom behavior analysis for enhanced engineering education: An AI-assisted approach. *International Journal of Computational Intelligence Systems, 17*, 167. https://doi.org/10.1007/s44196-024-00572-y
Hu, Q., & Rangwala, H. (2020). Towards fair educational data mining: A case study on detecting at-risk students. In A. N. Rafferty, J. Whitehill, V. Cavalli-Sforza, & C. Romero (Eds.), *Proceedings of The 13th international conference on educational data mining (EDM 2020), 2020* (pp. 431 – 437).
Huang, L. (2023). Ethics of artificial intelligence in education: Student privacy and data protection. *Science Insights Education Frontiers, 16*(2), 2577–2587. https://doi.org/10.15354/sief.23.re202
ISTE. (2023). *AI professional development for educators*. https://iste.org/ai
Karakuş, N., Gedik, K., & Kazazoğlu, S. (2025). Ethical decision-making in education: A comparative study of teachers and artificial intelligence in ethical dilemmas. *Behavioral Sciences, 15*(4), 469. https://doi.org/10.3390/bs15040469
Karataş, F., & Ataç, B. A. (2024). When TPACK meets artificial intelligence: Analyzing TPACK and AI-TPACK components through structural equation modelling. *Education and Information Technologies, 30*, 8979–9004. https://doi.org/10.1007/s10639-024-13164-2
Khan, S., Mazhar, T., Shahzad, T., Khan, M. A., Rehman, A. U., Saeed, M. M., & Hamam, H. (2025). Harnessing AI for sustainable higher education: Ethical considerations, operational efficiency, and future directions. *Discover Sustainability, 6*(1), 23. https://doi.org/10.1007/s43621-025-00809-6
Khanolkar, P. M., Vrolijk, A., & Olechowski, A. (2023). Mapping artificial intelligence-based methods to engineering design stages: A focused literature review. *AI EDAM, 37*, Article e25.
Koehler, M. J., & Mishra, P. (2009). What is technological pedagogical content knowledge (TPACK)? *Contemporary Issues in Technology and Teacher Education, 9*(1), 60–70.
Li, L., Chen, C. P., Wang, L., Liang, K., & Bao, W. (2023). Exploring artificial intelligence in smart education: Real-time classroom behavior analysis with embedded devices. *Sustainability, 15*(10), 7940. https://doi.org/10.3390/su15107940

Luckin, R., Holmes, W., Griffiths, M., & Forcier, L. B. (2016). *Intelligence unleashed: An argument for AI in education.* Pearson Education.

Melo-López, V. A., Basantes-Andrade, A., Gudiño-Mejía, C. B., & Hernández-Martínez, E. (2025). The impact of artificial intelligence on inclusive education: A systematic review. *Education Sciences, 15*(5), 539. https://doi.org/10.3390/educsci15050539

Microsoft. (2021). Accelerating learning analytics and AI in education. *Microsoft Education Report.*

Mishra, P., Warr, M., & Islam, R. (2023). TPACK in the age of ChatGPT and Generative AI. *Journal of Digital Learning in Teacher Education, 39*(4), 235–251. https://doi.org/10.1080/21532974.2023.2247480

Mollick, E. R., & Mollick, L. (2023). Using AI to implement effective teaching strategies in classrooms: Five strategies, including prompts. *The Wharton School Research Paper.* https://doi.org/10.2139/ssrn.4391243

Nguyen, A., Ngo, H. N., Hong, Y., Dang, B., & Nguyen, B. P. T. (2023). Ethical principles for artificial intelligence in education. *Education and Information Technologies, 28*, 4221–4241. https://doi.org/10.1007/s10639-022-11316-w

Ning, Y., Zhang, C., Xu, B., Zhou, Y., & Wijaya, T. T. (2024). Teachers' AI-TPACK: Exploring the relationship between knowledge elements. *Sustainability, 16*(3), 978. https://doi.org/10.3390/su16030978

Ruiz Viruel, S., Sánchez Rivas, E., & Ruiz Palmero, J. (2025). The role of artificial intelligence in project-based learning: Teacher perceptions and pedagogical implications. *Education Sciences, 15*(2), 150. https://doi.org/10.3390/educsci15020150

Salas-Pilco, S. Z., Xiao, K., & Hu, X. (2023). Artificial intelligence and learning analytics in teacher education: A systematic review. *Education Sciences, 12*(8), 569. https://doi.org/10.3390/educsci12080569

Selwyn, N. (2019). *Should robots replace teachers?* Polity Press.

Tan, J., & Yang, J. (2019). Ethical risks of artificial intelligence technology and collaborative governance. *Chinese Public Administration, 10*, 44–50. https://doi.org/10.19735/j.issn.1006-0863.2019.%2010.07

Touretzky, D. S., Gardner-McCune, C., Martin, F., & Seehorn, D. (2019). Envisioning AI for K–12: What should every child know about AI? *Proceedings of the AAAI Conference on Artificial Intelligence, 33*(01), 9795–9799. https://doi.org/10.1609/aaai.v33i01.33019795

Umar, M., Shaheen, S., Shehzad, T., ul Muntaha, S. S., & Khan, S. (2025). Teachers and technology: The evolving role of educators in aı-enhanced learning environments. *Indus Journal of Social Sciences, 3*(2), 600–616. https://doi.org/10.59075/ijss.v3i2.1431

UNESCO. (2021). *AI and education: Guidance for policy-makers.* https://unesdoc.unesco.org/ark:/48223/pf0000376709

UNESCO. (2024). *AI competency framework for teachers.* https://www.unesco.org/en/articles/ai-competency-framework-teachers

Williamson, B., & Eynon, R. (2020). Historical threads, missing links, and future directions in AI in education. *Learning, Media and Technology, 45*(3), 223–235. https://doi.org/10.1080/17439884.2020.1798995

World Economic Forum. (2024). *5 Key Policy Ideas to Integrate AI in Education Effectively.* https://www.weforum.org/stories/2024/04/prepare-future-policy-ideas-ai-in-education/

Xu, W., & Ouyang, F. (2022). The application of AI technologies in STEM education: A systematic review from 2011 to 2021. *International Journal of STEM Education, 9*(1), 59. https://doi.org/10.1186/s40594-022-00377-5

Zawacki-Richter, O., Marín, V. I., Bond, M., & Gouverneur, F. (2019). Systematic review of research on artificial intelligence applications in higher education – Where are the educators? *International Journal of Educational Technology in Higher Education, 16*(1), 1–27. https://doi.org/10.1186/s41239-019-0171-0

Zhai, X., Chu, X., Chai, C. S., Jong, M. S. Y., Istenic, A., Spector, M., & Li, Y. (2021). A Review of Artificial Intelligence (AI) in education from 2010 to 2020. *Complexity, 38*(12), 1182–1194. https://doi.org/10.1155/2021/8812542

Chapter 3
AI as a Teaching Tool: Empowering Pre-Service Science Teachers Through Conceptual Foundations and Classroom Integration

Tiffanie Zaugg

3.1 Introduction

The integration of artificial intelligence (AI) into educational contexts represents one of the most significant technological shifts in contemporary pedagogy (Adiguzel et al., 2023). As educational institutions worldwide grapple with the implications of digital transformation, science education stands at a particularly critical juncture. Science education's data-driven nature, inquiry-based learning emphasis, and reliance on computational tools position as both a natural beneficiary and testing ground for AI applications (Holmes et al., 2019).

Understanding AI's historical trajectory provides essential context for contemporary educational applications (Chiu et al., 2023). While AI might appear to be a recent phenomenon emerging suddenly in classrooms, its conceptual foundations reach back decades to foundational work by pioneers like Alan Turing and John McCarthy. This historical perspective helps educators appreciate AI's revolutionary potential and iterative nature, informing more thoughtful implementation strategies in education (Roll & Wylie, 2016).

This chapter examines the foundational elements necessary for understanding AI's role in science education, with particular attention to the historical development of AI technologies, the preparation of K-12 science teachers, the conceptual frameworks governing AI implementation, and the ethical considerations that must guide these

T. Zaugg (✉)
University of Central Florida, Central Florida, FL, USA
e-mail: tzaugg@centralriversaea.org

Central Rivers Area Education Agency, Cedar Falls, IA, USA

© The Author(s), under exclusive license to Springer Nature Switzerland AG 2026
M. Alkış Küçükaydın (ed.), *Preparing K-12 Pre-Service Science Teachers for the Age of Artificial Intelligence*, Advances in Technology-Rich Science Education 4,
https://doi.org/10.1007/978-3-032-18030-8_3

efforts (Baker & Hawn, 2022). The integration of practical frameworks and classroom applications demonstrates how theoretical foundations translate into actionable strategies for empowering science teachers through conceptual foundations and classroom integration.

3.2 Historical Foundations of AI

3.2.1 The Birth of AI as a Discipline

The idea of intelligent machines has existed in human imagination for centuries, but the field of artificial intelligence officially began in the mid-twentieth century. The term "artificial intelligence" was coined by computer scientist John McCarthy at the 1956 Dartmouth Summer Research Project on Artificial Intelligence. This pivotal conference gathered leading thinkers to explore creating machines that could simulate human intelligence, and this gathering is widely considered the birth of AI as an academic discipline (McCarthy et al., 1955).

Another key figure in AI's foundational period was British mathematician Alan Turing. In his seminal 1950 paper, "Computing Machinery and Intelligence," he posed the fundamental question, "Can machines think?" and proposed the Turing Test, a method to determine if a machine's behavior is indistinguishable from a human's. The Turing Test remains a foundational concept for evaluating machine intelligence and continues to influence contemporary discussions about AI capabilities in educational contexts (Turing, 1950).

3.2.2 From Turing to Generative AI: Key Milestones

Artificial intelligence emerged as a formal field in the mid-twentieth century, built on the work of Alan Turing and John McCarthy. Key milestones—from the Turing Test to modern generative systems—each advanced our understanding of machine learning, human–computer interaction, and their educational use (Roll & Wylie, 2016). The evolution from simple rule-based systems to sophisticated neural networks demonstrates AI's accelerating development and increasing relevance to educational applications.

3.2.3 Early Human–Computer Interaction: ELIZA and Its Educational Implications

In 1966, computer scientist Joseph Weizenbaum developed ELIZA, the first chatbot, to mimic a psychotherapist. The program used simple rules to rephrase user statements as questions, creating the illusion of understanding. This led to the ELIZA Effect, where users, including Weizenbaum's own secretary, projected human intelligence and emotions onto the program and confided in it (Weizenbaum, 1966). The ELIZA Effect is still relevant today, as students interact with AI tools like tutors and virtual labs (Jarrahi, 2018). This historical lesson highlights the need for AI literacy in science teacher preparation, so educators can teach students to critically evaluate AI outputs and distinguish between a program's actual capabilities and the illusion of understanding (Walter, 2024). While the ELIZA Effect demonstrated the pitfalls of over-attributing intelligence to machines, other work, such as Seymour Papert's (1980) *Mindstorms*, showed how computers could be tools for students to construct knowledge and foster agency.

3.2.4 Evolution Through Cycles of Innovation and Challenge

The history of AI is marked by periods of great excitement followed by periods of reduced funding and interest, often called "AI winters" (Roll & Wylie, 2016). Understanding these cycles helps educators maintain realistic expectations about AI capabilities while recognizing the technology's genuine potential for educational transformation.

3.2.5 1980s: The Revival of Expert Systems

This period saw the rise of expert systems—AI programs designed to emulate human expert decision-making in a specific domain. This era demonstrated AI's potential for domain-specific instruction and assessment, though the rigid rule-based nature of these systems limited their adaptability to diverse learning needs (Luckin et al., 2016).

3.2.6 1990s–2000s: Machine Learning and Strategic Thinking

The focus shifted from rule-based systems to machine learning, where computers learned from data rather than being explicitly programmed (Roll & Wylie, 2016).

3.2.7 2010s–Present: The Deep Learning Revolution

The 2010s marked a breakthrough with the rise of deep learning, a subfield of machine learning that uses multi-layered neural networks (Crompton & Burke, 2023). This advancement was fueled by the availability of vast amounts of data and more powerful computing hardware.

3.2.8 Generative AI Era: Current Educational Applications

The current era is defined by generative AI (e.g., ChatGPT, Google's Gemini), which creates new content like text, images, and code (Zhai, 2023). This technology is rapidly changing fields from creative arts to scientific research and education. For science educators, generative AI offers unprecedented opportunities for personalized learning materials, assessment item generation, and student inquiry support while raising important questions about academic integrity and authentic learning (Ahmad et al., 2024).

3.2.9 Implications for Science Education

This historical trajectory reveals several important considerations for contemporary science education:

Pattern of Overestimation and Realistic Assessment. The history of AI demonstrates a recurring pattern where initial excitement about capabilities leads to overestimation, followed by more realistic assessment (Roll & Wylie, 2016). Science educators must help students understand both the genuine capabilities and limitations of AI tools, using these instances to teach about the provisional and iterative nature of scientific understanding and models, emphasizing critical skepticism and evidence-based reasoning.

Evolution Toward Learning from Data. The shift from rule-based systems to machine learning approaches mirrors pedagogical movements toward data-driven instruction and evidence-based practice in education (Black & Wiliam, 2009).

Increasing Accessibility. While early AI systems required significant technical expertise, contemporary AI tools are increasingly accessible to educators and students, democratizing access to powerful computational capabilities (Chiu et al., 2023).

Human-AIInteraction Patterns. The ELIZA Effect and subsequent research on human–computer interaction provide crucial insights for designing educational experiences that leverage AI capabilities while maintaining critical thinking and human agency (Jarrahi, 2018). These considerations build on early visions of technology as a tool for student empowerment (Papert, 1980). Teachers can use this historical lesson to develop students' critical evaluation skills, helping them distinguish between genuine AI capabilities and the illusion of understanding when interacting with AI-powered scientific tools or generating content (Walter, 2024), thereby fostering essential scientific literacy (VanLehn, 2011; Vygotsky, 1978).

3.2.10 Preparing K-12 Science Teachers for the AI Age

3.2.10.1 The Current Landscape of Science Teacher Preparation

Preparation of science teachers has traditionally focused on content knowledge, pedagogical content knowledge, and general teaching methods (Shulman, 1986). Foundational documents such as the National Research Council's (2012) *A Framework for K-12Science Education: Practices, Crosscutting Concepts, and Core Ideas*, which outline key scientific practices, crosscutting concepts, and core ideas essential for K-12 science learning further contextualize this content knowledge. These competencies must enable them to implement the Next Generation Science Standards (NGSS Lead States, 2013), which emphasize three-dimensional learning, inquiry, and scientific practices, within AI-enhanced environments. However, the emergence of AI as a transformative force in education necessitates a fundamental reconsideration of teacher preparation programs (Ng et al., 2023). Contemporary teachers must be equipped not only with traditional competencies but also with the TPACK necessary to effectively integrate AI tools into their instructional practices (Mishra & Koehler, 2006).

3.2.10.2 TPACK

Successful preparation relies on the TPACK model—integrating technological, pedagogical, and content knowledge in designing instruction (Mishra & Koehler, 2006). For AI integration, teachers must weave together expertise from all three domains, leveraging practical classroom examples for authentic learning.

The historical development of AI from the 1950s Dartmouth Conference to contemporary generative AI systems illustrates both the rapid pace of technological advancement and the importance of understanding fundamental principles that persist across technological generations. Teachers need this historical context to appreciate that AI integration is not merely about adopting the latest tools but about understanding how artificial intelligence can support fundamental pedagogical goals (Holmes et al., 2019). This deeper "Technology Knowledge" component of TPACK (Mishra & Koehler, 2006) is crucial for avoiding the pitfalls of over-attribution of intelligence, as demonstrated by the ELIZA Effect (Weizenbaum, 1966), and for discerning genuine AI capabilities from illusions of understanding (Walter, 2024).

Research indicates many pre-service science teachers enter their programs with limited understanding of AI technologies and their potential applications in educational contexts (García-Peñalvo & Corell, 2020). This knowledge gap is particularly concerning given these future educators will be responsible for preparing students for a world increasingly shaped by artificial intelligence. The challenge is compounded by the rapid pace of technological change, which often outpaces the ability of teacher preparation programs to adapt their curricula accordingly (Bayne et al., 2023).

3.2.11 Practical TPACK Example

A chemistry teacher uses AI modeling software to teach chemical bonding concepts. This integration demonstrates the intersection of all three TPACK domains (Mishra & Koehler, 2006):

- *Content Knowledge*: Understanding of molecular bonding, electron configurations, and chemical interactions as outlined by the disciplinary core ideas within the National Research Council's (2012) A Framework for K-12 Science Education.
- *Pedagogical Knowledge*: Recognizing common student misconceptions about abstract concepts and using inquiry-based learning approaches (Krajcik & Shin, 2014) that align with the scientific practices outlined in the Next Generation Science Standards (NGSS Lead States, 2013).
- *Technology Knowledge*: Understanding how AI systems generate molecular models, their limitations in representing quantum phenomena, and interpreting software outputs. This includes comprehending the underlying machine learning processes (Zhai & Nehm, 2024) and recognizing potential biases or errors, ensuring the teacher can guide students in a critical evaluation of AI-generated content.

The teacher designs inquiry-based labs accessible to diverse learners, where students use AI-powered molecular modeling to explore chemical bonding while developing scientifically accurate mental models. This approach leverages the AI system's visualization capabilities while maintaining student agency in learning (Merchant et al., 2014).

3.2.12 Digital Transformation in Science Education Through UDL Principles

Digital transformation in science education extends beyond the mere adoption of new technologies; it represents a fundamental shift in how scientific concepts are taught, learned, and understood (Pelgrum & Law, 2003). AI technologies offer unique advantages for science education, including the ability to simulate complex phenomena, analyze large datasets, and provide personalized feedback to learners (Rutten et al., 2012). However, realizing these benefits requires careful consideration of how AI tools align with established principles of science pedagogy and accessibility frameworks such as UDL.

UDL is based on research in cognitive neuroscience that guides the development of flexible learning environments that can accommodate learner variability (Novak & Rodriguez, 2023). The three principles of UDL—multiple means of engagement, representation, and action and expression—provide a robust framework for ensuring that AI integration serves all learners effectively in science education contexts (CAST, 2024).

3.2.13 UDL Framework

UDL offers three actionable pathways that align seamlessly with AI capabilities (Saborío-Taylor & Rojas-Ramírez, 2024):

- *Engagement*: Learning connects students' interests, backgrounds, and motivational needs. AI tools can support engagement by connecting science learning to students'authentic interests and cultural backgrounds (Peng et al., 2024). For example, an environmental science teacher might use AI-powered climate modeling tools that allow students to investigate environmental issues affecting their local communities. Students from coastal regions might explore sea-level rise impacts, while those in agricultural areas might investigate changing precipitation patterns. The AI system can provide real-time data and predictive models that make abstract climate concepts tangible and personally relevant.
- *Representation*: Multiple ways (visual, auditory, tactile) to present scientific material, ensuring access for students with disabilities (Hyatt & Owenz, 2024). AI technologies excel at providing multiple representations of scientific phenomena. A physics teacher studying wave behavior might use AI-powered visualization tools that simultaneously present wave phenomena through mathematical equations, graphical representations, animated simulations, and real-world examples. Students with visual processing differences might benefit from audio descriptions generated by AI text-to-speech systems, while students who are deaf or hard of hearing might access vibrotactile representations of wave phenomena through AI-controlled haptic devices.
- *Action & Expression*: Flexible options for students to show understanding—written explanations, AI-narrated videos, digital posters, etc. AI tools can provide multiple pathways for students to demonstrate their scientific understanding (Edutopia, 2024). A chemistry teacher assessing students' understanding of molecular interactions might allow students to choose between traditional written explanations, AI-assisted video presentations where students narrate molecular

animations, collaborative digital posters created with AI design tools, or hands-on modeling using AI-guided 3D printing to create physical molecular models.

AI integration in science education must be grounded in constructivist learning theories emphasizing student-centered inquiry and knowledge construction (Vygotsky, 1978). AI tools can support these approaches by providing scaffolding for complex investigations, enabling students to explore phenomena that would otherwise be inaccessible, and facilitating collaborative learning experiences (Collins et al., 1989). For example, AI-powered virtual laboratories can allow students to conduct experiments that would be too dangerous, expensive, or time-consuming to perform in traditional settings (de Jong et al., 2013).

3.2.14 Building AI Literacy and Ethical Capacity

3.2.14.1 Core AI Competencies

Preparing science teachers for the AI age requires the development of specific competencies that bridge technological proficiency with pedagogical expertise and UDL principles (Long & Magerko, 2020). These competencies include understanding AI fundamentals, evaluating AI tools for educational appropriateness, designing AI-enhanced learning experiences, and addressing ethical considerations in AI implementation (Crompton & Burke, 2023). While foundational, it is essential for science educators to apply these competencies within the context of scientific practices and K-12 learning environments (Zhai & Nehm, 2024).

- *Machine Learning Fundamentals*: Understanding how data trains systems and their applications in science education contexts, such as species identification or image recognition (Zhai & Nehm, 2024). Teachers must comprehend how algorithms learn from data and make predictions. A biology teacher working with AI-powered species identification tools should understand that the system has been trained on thousands of images and can recognize patterns in visual features like leaf shape, flower structure, or animal markings. This knowledge enables the teacher to help students understand why the AI might misidentify a rare species or struggle with poor-quality images, fostering critical evaluation of AI outputs. This understanding is crucial for K-12 science teachers to critically evaluate and integrate AI tools that support scientific inquiry and learning, aligning with diverse learners' specific needs (Zhai & Nehm, 2024).
- *Natural Language Processing*: Text and language interpretation capabilities, with a focus on potential output errors or bias (Walter, 2024). Science teachers need to understand how AI systems interpret and generate text. An earth science teacher using AI-powered research assistance tools should recognize that these systems can summarize scientific literature, identify key concepts, and suggest research connections. However, they must also understand limitations such as the AI's potential to miss nuanced arguments, misinterpret technical terminology,

or generate plausible sounding but incorrect information. This understanding is crucial for K-12 science teachers to critically evaluate and integrate AI tools that support scientific inquiry and learning, aligning with diverse learners' specific needs (Zhai & Nehm, 2024).

- *Data Analytics & Visualization*: Large dataset analysis capabilities, supporting student investigations and personalized feedback (Siemens, 2013). Teachers must comprehend how AI systems process and analyze large datasets. A chemistry teacher implementing AI-powered spectroscopy analysis should understand how machine learning algorithms identify compounds by comparing spectral patterns to reference databases. This knowledge allows the teacher to help students appreciate AI's power in handling complex analytical tasks and the importance of high-quality reference data for accurate results. This understanding is crucial for K-12 science teachers to critically evaluate and integrate AI tools that support scientific inquiry and learning, aligning with aligning with diverse learners' specific needs he specific needs of diverse learners (Zhai & Nehm, 2024).
- *Computer Vision*: Tools for video analysis or automated observation in experiments (Kim et al., 2022). Science teachers should understand how AI systems interpret visual information. A physics teacher using AI-powered motion analysis software should comprehend how the system tracks objects in video, calculates velocities and accelerations, and generates motion graphs. This understanding enables the teacher to help students recognize when environmental factors (lighting, contrast, occlusion) might affect AI accuracy and develop strategies for optimizing experimental setups. This understanding is crucial for K-12 science teachers to critically evaluate and integrate AI tools that support scientific inquiry and learning, aligning with diverse learners' specific needs (Zhai & Nehm, 2024).

3.2.15 Teacher Tasks and Competencies

Critical Evaluation Skills:

- Critically evaluate AI tool outputs for bias (Baker & Hawn, 2022).
- Interpret analytics to support individual student needs (Jivet et al., 2024).

3.2.15.1 Pedagogical Integration Competencies

Building on foundational AI literacy, teachers must develop practical pedagogical competencies for effectively integrating AI into the science classroom. These competencies include:

- *Tool Selection and Evaluation*: Teachers need skills to evaluate AI tools for pedagogical appropriateness, accuracy, and alignment with learning standards and

UDL principles (Scherer et al., 2019). For instance, a life science teacher considering AI-powered virtual dissection software would assess its anatomical accuracy, support for inquiry-based exploration, alignment with comparative anatomy standards, and scaffolding for diverse learners.

- *Learning Experience Design*: Teachers must design instructional sequences that meaningfully integrate AI tools as core components rather than mere add-ons, incorporating UDL principles throughout (Holstein et al., 2019). For example, an environmental science teacher could design a multi-week investigation where students use AI-powered satellite image analysis to track deforestation patterns, connect their findings to climate data using AI analytics tools, and develop conservation proposals informed by AI-generated predictive models.

3.2.16 Assessment and Evaluation Competencies

- *AI-Enhanced Formative Assessment*: Teachers must learn to use AI tools to gather real-time information about student understanding and adjust instruction accordingly, while ensuring that assessment methods are accessible to all learners (Black & Wiliam, 2009). A biology teacher might use AI-powered analysis of student drawings of cell structures to quickly identify common misconceptions, and then provide targeted mini-lessons addressing specific conceptual gaps.
- *Authentic Performance Assessment Design*: Teachers must develop skills in creating assessments that require genuine understanding rather than AI-assisted task completion, while providing multiple pathways for demonstrating mastery (Kim et al., 2022). This includes designing assessments that allow students to demonstrate understanding across the three dimensions of the Next Generation Science Standards (NGSS Lead States, 2013), especially with AI-enabled scoring, aligning with scientific practices and crosscutting concepts articulated in the National Research Council's (2012) A Framework for K-12 Science Education.

3.2.17 Ethics Checklist for Classroom Use

- *Data Privacy and Security Management*: Teachers must understand how to protect student data when using AI systems, with particular attention to the privacy rights of students with disabilities and those from vulnerable populations (Reidenberg & Schaub, 2018). An environmental science teacher implementing AI-powered citizen science platforms should know how to configure privacy settings, understand data sharing agreements, obtain appropriate permissions, and educate students about digital privacy.
- *Algorithmic Bias Recognition and Mitigation*: Teachers must develop skills in identifying and addressing bias in AI systems, particularly bias that might affect students with disabilities or from underrepresented groups (Baker & Hawn, 2022).

A biology teacher using AI-powered disease diagnosis simulations should recognize that training data might under-represent certain populations, leading to biased diagnostic suggestions.

3.3 Implementation Strategies: SAMR & UDL in Action

3.3.1 Progression of AI Integration

The SAMR model provides a useful framework for thinking about technology integration depth (Puentedura, 2006), while UDL principles ensure that integration efforts serve all learners effectively (Hamilton et al., 2016).

- *Substitution*: Digital drills with flexible input options. At this level, an AI-powered digital worksheet might replace traditional paper-based practice problems in chemistry, where students input chemical formulas and receive immediate feedback on their accuracy. To align with UDL principles, the digital worksheets might offer multiple input methods (typing, voice recognition, stylus input) and provide feedback in multiple formats (text, audio, visual indicators).
- *Augmentation*: Adaptive hints, multimodal explanations. The AI tool begins to add functional improvements to the traditional task while expanding accessibility options (Hamilton et al., 2016). For example, an AI chemistry tutor might not only provide correct answers but also offer adaptive hints based on common student errors, track performance over time, and adjust problem difficulty based on student mastery.
- *Modification*: Open-ended, inquiry-rich simulations highlighting accessibility. This level involves significant task redesign enabled by AI capabilities, with accessibility considerations fundamentally transforming the learning experience (Peng et al., 2024). Consider an AI-powered environmental science simulation where students investigate climate change by manipulating multiple variables (carbon emissions, deforestation rates, renewable energy adoption) and observing long-term environmental impacts through accelerated time modeling.
- *Redefinition*: Real-time feedback, multi-modal collaboration, new authentic tasks beyond traditional means. AI enables entirely new forms of learning that would be inconceivable without the technology, while UDL principles ensure these innovations serve all learners (Chiu, 2025). An example might be an AI system that analyzes students' verbal explanations of scientific phenomena, identifies conceptual gaps in real-time, and generates personalized learning pathways that connect students with peer experts, relevant research articles, virtual field experiences, and adaptive assessments (Table 3.1).

Table 3.1 Example implementation table: AI integration in science classrooms

Level	AI use case	UDL features
Substitution	Worksheet auto-grading	Text/voice input, flexible output
Augmentation	Chemistry tutor with adaptive hints	Peer video, animated guides
Modification	Climate simulation for inquiry projects	Voice, gesture, tactile control
Redefinition	AI feedback on verbal science responses	Expression via speech, sign, image

3.4 Case Studies and Classroom Practice

3.4.1 Example 1: Accessible Virtual Laboratories

AI-powered labs offer accessible variable manipulation and provide options for students with motor or vision impairments through eye-tracking, audio feedback, haptics, and flexible recording modes (de Jong et al., 2013). These environments allow students to manipulate variables, observe outcomes, and develop scientific reasoning skills in ways that complement traditional laboratory experiences.

A virtual chemistry laboratory investigating reaction kinetics might allow students to manipulate temperature and concentration variables through multiple input methods. Students with motor impairments might use eye-tracking technology to select and move virtual equipment, while students with visual processing differences might rely on audio descriptions of color changes and receive tactile feedback about temperature variations through haptic devices (Hyatt & Owenz, 2024).

3.4.2 Example 2: Physics Simulations

Multi-modal representations of complex concepts through animations, audio, and tactile experiences ensure students can explore scientific phenomena according to their learning strengths (Merchant et al., 2014). A quantum mechanics simulation might represent wave-particle duality through multiple simultaneous representations: dynamic visual animations for students who learn best through visual-spatial processing, audio frequency changes that correspond to energy levels for students who process auditory information effectively, and haptic vibrations that correspond to wave amplitude for students who benefit from tactile input.

3.4.3 Example 3: Authentic Assessment

AI enables flexible performance assessments, allowing students to demonstrate scientific reasoning through reports, presentations, digital maps, and hands-on projects

(Nehm et al., 2012). A comprehensive chemistry assessment might ask students to investigate the effectiveness of different antacids using both AI-powered pH analysis tools and traditional laboratory techniques. Students demonstrate their understanding through their choice of multiple formats: detailed laboratory reports, video demonstrations with explanatory narration, interactive digital presentations, or collaborative poster sessions.

3.5 AI As a Teaching Tool: Conceptual Framework and Classroom Practices

3.5.1 Theoretical Foundations

The effective integration of AI into science education requires a robust theoretical framework that addresses both the technological and pedagogical dimensions of implementation while ensuring accessibility and inclusion for all learners (Bitzenbauer et al., 2024). TAM provides valuable insights into the factors that influence teacher adoption of AI technologies, including perceived usefulness and ease of use (Davis, 1989).

3.5.2 TAM in Practice

TAM's application to AI in science education can be illustrated through a case study of physics teachers adopting AI-powered virtual physics laboratories (Scherer et al., 2019). According to TAM, teachers' intention to use these tools depends primarily on two factors:

- *Perceived Usefulness:* The AI lab's ability to simulate expensive or dangerous experiments (such as radioactivity studies or high-energy particle interactions), provide immediate feedback on student hypotheses, and generate real-time data visualizations that would be impossible in traditional laboratory settings.
- Perceived Ease of Use: Factors such as AI interface intuitiveness, technical support quality, and time for teacher proficiency (Davis, 1989).

3.5.3 Types of AI Tools in Science Education and UDL Applications

- *Intelligent Tutoring Systems*: Provide personalized instruction and feedback, adapting to individual student needs and learning patterns (VanLehn, 2011). From a UDL perspective, effective intelligent tutoring systems offer adapted

content, provide content through various modalities (text, audio, visual, interactive simulations), and allows students to respond through different input methods.
- *Automated Assessment Tools*: Evaluate student responses to open-ended questions, provide immediate feedback, and identify common misconceptions (Nehm et al., 2012). When AI assessment tools provide multiple options for students to demonstrate their understanding—such as allowing written responses, oral explanations, visual diagrams, or multimedia presentations, students will consistently choose the modalities that best match their learning preferences and abilities.
- *Data Analytics and Learning AnalyticsPlatforms*: Enable educators to gain insights into student learning patterns, identify at-risk learners, and make data-driven instructional decisions (Siemens, 2013). When designed with UDL, these analytics platforms can provide insights into how different students prefer to access information, engage with content, and express their understanding (Jivet et al., 2024).

3.6 Reflection, Self-Assessment, and Discussion

3.6.1 Reflection Prompts

- *Lesson Planning Integration*: Draft a lesson plan incorporating AI and UDL for a science topic, identifying specific accessibility features and multiple pathways for student engagement (Garrison et al., 2000). Consider how AI's historical development affects current classroom practice and how AI tools can encourage inquiry and agency for students with different needs.
- *Critical Evaluation*: What risks might automated assessment pose to equity, and how can teachers mitigate them (Baker & Hawn, 2022)? How can the ELIZA Effect inform our approach to developing students' critical evaluation skills when interacting with AI systems?

3.7 Inclusive AI Learning Design Framework

Inclusive AI Learning Design Framework is directly based on UDL principles, integrates the 'why,' 'what,' and 'how' of learning with AI capabilities, maintaining a strong focus on accessibility and inclusion for all learners (Peng et al., 2024). This framework ensures AI tools enhance engagement by providing personalized connections to students' interests and backgrounds, offer multiple means of representation through diverse modalities for complex scientific concepts, and support multiple means of action and expression by offering varied pathways for students to interact with content and demonstrate understanding.

3.8 Conclusion

3.8.1 Foundations for Equitable AI-Enhanced Science Education

Effective teacher preparation for AI integration must center accessibility, equity, ethical practice, and learner agency (Williamson et al., 2020). By leveraging conceptual models (TPACK, UDL, SAMR, TAM) and critically applying AI tools, teachers can build science classrooms where technology prioritizes inclusion and authentic inquiry (Woods, 2024). Understanding this history will help teachers guide their students in using AI tools responsibly and prepare them for a future where AI is an integral part of science and society.

The integration of artificial intelligence into science education represents both an unprecedented opportunity and a significant challenge for educators, students, and educational institutions (Chiu et al., 2023). The historical perspective reveals that AI integration is not merely about adopting the latest technological innovations but about understanding fundamental principles of human–computer interaction that have persisted across decades of development. From the early insights of the ELIZA Effect to contemporary challenges with generative AI, educators must help students develop critical evaluation skills that distinguish between genuine AI capabilities and the illusion of understanding (Walter, 2024).

Preparing K-12 science teachers for the AI age demands a fundamental reconceptualization of teacher education programs that places accessibility and inclusion at the center of technological integration efforts (Bayne et al., 2023). These programs must equip future educators with the technological pedagogical content knowledge necessary to effectively integrate AI tools while maintaining fidelity to established principles of science pedagogy and universal design (Ng et al., 2023).

The integration of UDL principles throughout AI implementation ensures that technological advancement translates to educational equity rather than increased barriers for students with diverse needs (Saborío-Taylor & Rojas-Ramírez, 2024). By embedding UDL's framework, AI tools can enhance learning opportunities for all students, providing varied pathways for engagement, representation, and expression.

Perhaps most importantly, the ethical dimensions of AI implementation in science education cannot be treated as an afterthought or separate consideration (Baker & Hawn, 2022). Issues of privacy, bias, transparency, and student agency must be central considerations in any AI integration effort, with particular attention to how these issues affect students from diverse backgrounds and with varying abilities (Jobin et al., 2019).

The convergence of AI technologies with accessibility frameworks like UDL represents a powerful opportunity to address longstanding challenges in science education while preparing students for a future where human-AI collaboration will be essential for scientific advancement (Huang et al., 2024). By centering accessibility, equity, and learner agency in AI implementation efforts, educators can ensure that technological progress supports rather than undermines the fundamental goal of science education: developing scientifically literate citizens who can think critically, solve problems creatively, and participate meaningfully in democratic society regardless of their diverse abilities, backgrounds, and ways of learning.

3.8.2 *Future Research Directions in AI-Enhanced Science Education*

As artificial intelligence continues to evolve, ongoing research and reflective practice remain crucial to ensure its equitable and effective integration into science education (Zhai & Nehm, 2024). Future inquiry should prioritize several key areas. Firstly, longitudinal studies are needed to examine the long-term impact of AI implementation across diverse student populations, paying particular attention to how AI affects learning outcomes, engagement, and the development of scientific literacy for students with varying abilities and backgrounds. This would involve a continuous focus on issues of equity and access in AI-powered educational tools, building on considerations for the datafication of teaching (Williamson et al., 2020). Secondly, there is a critical need for developing new assessment methods that can effectively capture learning complexity in AI-enhanced environments (Zhai, 2024). These methods must accommodate diverse forms of expression while accurately evaluating students' scientific reasoning and understanding in an AI-integrated context (Kim et al., 2022). Finally, continued attention to the ethical dimensions of AI implementation is paramount, including in-depth research into mitigating algorithmic bias (Baker & Hawn, 2022) and safeguarding data privacy (Reidenberg & Schaub, 2018) within new AI tools and platforms. By fostering these research avenues, educators can ensure that AI technological progress genuinely supports rather than undermines the fundamental goals of science education.

References

Adiguzel, T., Kaya, M. H., & Cansu, F. K. (2023). Revolutionizing education with AI: Exploring the transformative potential of ChatGPT. *Contemporary Educational Technology, 15*(3), ep429.

Ahmad, S. F., et al. (2024). Evolution to revolution: Critical exploration of educators' perceptions of the impact of Artificial Intelligence (AI) on the teaching and learning process in the GCC region. *Computers and Education: Artificial Intelligence, 6*, 100132.

Baker, R. S., & Hawn, A. (2022). Algorithmic bias in education. *International Journal of Artificial Intelligence in Education, 32*(4), 1052–1092.

Bandura, A. (2006). Guide for constructing self-efficacy scales. In F. Pajares & T. Urdan (Eds.), *Self-efficacy beliefs of adolescents* (pp. 307–337). Information Age Publishing.

Bayne, S., Bali, M., & Dolch, C. (2023). Do we still need teachers? Navigating the paradigm shift of the teacher's role in the AI era. *Frontiers in Education, 8*, 1161777.

Bitzenbauer, P., et al. (2024). Human and AI collaboration in the higher education environment: Opportunities and concerns. *Cognitive Research: Principles and Implications, 9*(1), 1–29.

Black, P., & Wiliam, D. (2009). Developing the theory of formative assessment. *Educational Assessment, Evaluation and Accountability, 21*(1), 5–21.

CAST. (2024). *The UDL guidelines*. Retrieved from https://udlguidelines.cast.org/

Chiu, T. K. F. (2025). Encouraging human-AI collaboration in interactive learning environments. *Interactive Learning Environments, 33*(2), 1–3.

Chiu, T. K. F., Xia, Q., Zhou, X., Chai, C. S., & Cheng, M. (2023). Systematic literature review on opportunities, challenges, and future research recommendations of artificial intelligence in education. *Computers and Education: Artificial Intelligence, 4*, 100118.

Collins, A., Brown, J. S., & Newman, S. E. (1989). Cognitive apprenticeship: Teaching the crafts of reading, writing, and mathematics. In L. B. Resnick (Ed.), *Knowing, Learning, and Instruction* (pp. 453–494). Lawrence Erlbaum Associates.

Crompton, H., & Burke, D. (2023). Artificial intelligence in higher education: The state of the field. *Computers & Education, 194*, 104727.

Davis, F. D. (1989). Perceived usefulness, perceived ease of use, and user acceptance of information technology. *MIS Quarterly, 13*(3), 319–340.

de Jong, T., Linn, M. C., & Zacharia, Z. C. (2013). Physical and virtual laboratories in science and engineering education. *Science, 340*(6130), 305–308.

Edutopia. (2024). *Using AI in universal design for learning*. Retrieved from https://www.edutopia.org/article/using-ai-in-universal-design-for-learning/

García-Peñalvo, F. J., & Corell, A. (2020). The COVID-19: The enzyme of the digital transformation of teaching or the reflection of a methodological and competence crisis in higher education? *Campus Virtuales, 9*(2), 83–98.

Garrison, D. R., Anderson, T., & Archer, W. (2000). Critical inquiry in a text-based environment: Computer conferencing in higher education. *The Internet and Higher Education, 2*(2–3), 87–105.

Hamilton, E. R., Rosenberg, J. M., & Akcaoglu, M. (2016). The substitution augmentation modification redefinition (SAMR) model: A critical review and suggestions for its use. *TechTrends, 60*(5), 433–441.

Holmes, W., Bialik, M., & Fadel, C. (2019). *Artificial intelligence in education: Promises and implications for teaching and learning*. Center for Curriculum Redesign.

Holstein, K., McLaren, B. M., & Aleven, V. (2019). Co-designing a real-time classroom orchestration tool to support teacher–AI complementarity. *Journal of Learning Analytics, 6*(2), 27–52.

Huang, X., et al. (2024). Examining human–AI collaboration in hybrid intelligence learning environments: Insight from the Synergy Degree Model. *Humanities and Social Sciences Communications, 11*(1), 1–15.

Hyatt, S. E., & Owenz, M. B. (2024). Using universal design for learning and artificial intelligence to support students with disabilities. *College Teaching*, 1–12.

Jarrahi, M. H. (2018). Artificial intelligence and the future of work: Human-AI symbiosis in organizational decision making. *Business Horizons, 61*(4), 577–586.

Jivet, I., Wong, J., Scheffel, M., Valle Torre, M., Specht, M., & Drachsler, H. (2024). Learning analytics and the Universal Design for Learning (UDL): A clustering approach. *Computers & Education, 208*, 104960.

Jobin, A., Ienca, M., & Vayena, E. (2019). The global landscape of AI ethics guidelines. *Nature Machine Intelligence, 1*(9), 389–399.

Kim, Y., et al. (2022). Rubric development for AI-enabled scoring of three-dimensional constructed-response assessment aligned to NGSS learning progression. *Frontiers in Education, 7*, 983055.

Krajcik, J. S., & Shin, N. (2014). Project-based learning. In R. K. Sawyer (Ed.), *The Cambridge handbook of the learning sciences* (2nd ed., pp. 275–297). Cambridge University Press.

Long, D., & Magerko, B. (2020). What is AI literacy? Competencies and design considerations. In *Proceedings of the 2020 CHI Conference on Human Factors in Computing Systems* (pp. 1–16).

Luckin, R., Holmes, W., Griffiths, M., & Forcier, L. B. (2016). *Intelligence unleashed: An argument for AI in education*. Pearson.

McCarthy, J., Minsky, M. L., Rochester, N., & Shannon, C. E. (1955). A proposal for the Dartmouth summer research project on artificial intelligence.

Merchant, Z., Goetz, E. T., Cifuentes, L., Keeney-Kennicutt, W., & Davis, T. J. (2014). Effectiveness of virtual reality-based instruction on students' learning outcomes in K-12 and higher education: A meta-analysis. *Computers & Education, 70*, 29–40.

Mishra, P., & Koehler, M. J. (2006). Technological pedagogical content knowledge: A framework for teacher knowledge. *Teachers College Record, 108*(6), 1017–1054.

National Research Council. (2012). *A framework for K-12 science education: Practices, crosscutting concepts, and core ideas*. The National Academies Press.

Nehm, R. H., Urban-Lurain, M., & Rowell, G. H. (2012). Assessment and measurement of conceptual knowledge in the biological sciences: A cross-disciplinary approach for concept inventory development. *CBE—Life Sciences Education, 11*(4), 413–424.

Ng, D. T. K., Leung, J. K. L., Su, J., Ng, R. C. W., & Chu, S. K. W. (2023). Teachers' AI digital competencies and twenty-first century skills in the post-pandemic world. *Educational Technology Research and Development, 71*(1), 137–161.

NGSS Lead States. (2013). *Next generation science standards: For states*. The National Academies Press.

Novak, K., & Rodriguez, K. (2023). *In support of students: A leader's guide to equitable MTSS*. Wiley.

NSTA. (2024). *Empowering science education with AI: Enhancing three-dimensional learning*. Retrieved from https://www.nsta.org/blog/empowering-science-education-aienhancing-three-dimensional-learning

Papert, S. (1980). *Mindstorms: Children, computers, and powerful ideas*. Basic Books.

Pelgrum, W. J., & Law, N. (2003). *ICT in education around the world: Trends, problems and prospects*. UNESCO International Institute for Educational Planning.

Peng, C., Chen, S., Tsai, M. J., Weisberg, L. R., & Tsai, C. C. (2024). A framework for inclusive AI learning design for diverse learners. *Computers and Education Open, 5*, 100134.

Puentedura, R. R. (2006). *Transformation, technology, and education*. Hippasus. http://hippasus.com/resources/tte/

Reidenberg, J. R., & Schaub, F. (2018). Achieving big data privacy in education. *Theory and Research in Education, 16*(3), 263–279.

Roll, I., & Wylie, R. (2016). Evolution and revolution in artificial intelligence in education. *International Journal of Artificial Intelligence in Education, 26*(2), 582–599.

Rutten, N., van Joolingen, W. R., & van der Veen, J. T. (2012). The learning effects of computer simulations in science education. *Computers & Education, 58*(1), 136–153.

Saborío-Taylor, S., & Rojas-Ramírez, F. (2024). Universal design for learning and artificial intelligence in the digital era: Fostering inclusion and autonomous learning. *International Journal of Professional Development, Learners and Learning, 6*(2), ep2408.

Scherer, R., Siddiq, F., & Howard, S. K. (2019). Technology acceptance and technology use: Applying the TAM to understand teachers' technology use in 21st century classrooms. *Educational Technology Research and Development, 67*(4), 859–881.

Shulman, L. (1986). Those who understand: Knowledge growth in teaching. *Educational Researcher, 15*(2), 4–14.

Siemens, G. (2013). Learning analytics: The emergence of a discipline. *American Behavioral Scientist, 57*(10), 1380–1400.

Teacher to Techie. (2025). *AI for science teachers: Top 10 tools*. Retrieved from https://teachertotechie.org/ai-for-science-teachers-top-10-tools/

Turing, A. M. (1950). Computing machinery and intelligence. *Mind, 59*(236), 433–460.

VanLehn, K. (2011). The relative effectiveness of human tutoring, intelligent tutoring systems, and other tutoring systems. *Educational Psychologist, 46*(4), 197–221.

Vygotsky, L. S. (1978). *Mind in society: The development of higher psychological processes*. Harvard University Press.

Walter, Y. (2024). Embracing the future of Artificial Intelligence in the classroom: The relevance of AI literacy, prompt engineering, and critical thinking in modern education. *International Journal of Educational Technology in Higher Education, 21*, 15.

Weizenbaum, J. (1966). ELIZA—A computer program for the study of natural language communication between man and machine. *Communications of the ACM, 9*(1), 36–45.

Williamson, B., Bayne, S., & Shay, S. (2020). The datafication of teaching in higher education: Critical issues and perspectives. *Teaching in Higher Education, 25*(4), 351–365.

Woods, G. (2024). *The AI-driven leader*. Published September 16, 2024.

Zhai, X. (2023). ChatGPT user experience: Implications for education. *SSRN Electronic Journal*.

Zhai, X. (2024). AI and machine learning for next generation science assessments. arXiv preprint arXiv:2405.06660

Zhai, X., & Nehm, R. H. (2024). The impact of artificial intelligence on scientific practices: An emergent area of research for science education. *International Journal of Science Education, 46*(4), 361–384.

Chapter 4
AI Integration in Science Learning: Mapping K-12 Pre-Service Teachers' Competencies, Pedagogical Practices, and Simulation-Based Learning Approaches

Iwan Wicaksono

4.1 Introduction

The learning paradigm is undergoing a fundamental shift from traditional methods toward interactive, collaborative, and technology-enhanced approaches. The integration of digital devices broadens access to information, enriches learning experiences, and deepens students' contextual understanding of scientific concepts. In this transformation, Artificial Intelligence (AI) serves as a strategic catalyst for innovation in science education. AI facilitates adaptive, personalized, and data-driven learning environments that accommodate diverse student needs, while still requiring the presence of pre-service teachers as facilitators who provide pedagogical guidance and feedback (Kim et al., 2022).

Pre-service K-12 teachers remain central to bridging technological integration with the pedagogical objectives of science education. Their role is indispensable in ensuring that technology use aligns with the development of knowledge, critical thinking, and scientific attitudes. Effective engagement with AI demands that pre-service teachers possess strong digital literacy and technology-based pedagogical competencies. These skills enable them to leverage innovation, design context-appropriate learning strategies, and address implementation challenges in digital learning environments (Koh et al., 2023).

Beyond serving as a learning medium, AI also supports instructional design and assessment practices. Its application allows pre-service teachers to develop more effective teaching methods, monitor student progress in real time, and enhance

I. Wicaksono (✉)
University of Jember, Jember, Indonesia
e-mail: iwanwicaksono.fkip@unej.ac.id

© The Author(s), under exclusive license to Springer Nature Switzerland AG 2026

M. Alkış Küçükaydın (ed.), *Preparing K-12 Pre-Service Science Teachers for the Age of Artificial Intelligence*, Advances in Technology-Rich Science Education 4,
https://doi.org/10.1007/978-3-032-18030-8_4

assessment quality through adaptive automated evaluation systems (Puspita et al., 2023). Consequently, AI contributes to improving both the processes and outcomes of science learning. However, key challenges persist, particularly regarding pre-service teachers' AI literacy, readiness for AI-based instruction, and attitudes toward technological adoption. Strengthening AI literacy among pre-service K-12 teachers is therefore essential to ensuring the effective and sustainable integration of AI in science education.

4.2 Background

Learning in the 21st century is being reshaped by the Industrial Revolution 4.0 and ongoing digital transformation, which have altered social, economic, and communicative paradigms and ushered education into the era of digital pedagogy. However, research indicates that most pre-service teachers lack sufficient conceptual and practical competence in artificial intelligence (AI), highlighting the need for comprehensive professional development strategies (Bustanul Arifin & Abdul Mu'id, 2024).

AI-based simulations extend beyond virtual experimentation by presenting complex scenarios that foster critical thinking and decision-making (Amrullah et al., 2025). They function as cognitive scaffolds, enhancing analytical and evaluative thinking through authentic, problem-oriented learning experiences that strengthen conceptual understanding and reflective reasoning (Koh et al., 2023).

Developing analytical and problem-solving skills is essential for connecting theoretical knowledge to real-world contexts and addressing scientific and everyday challenges. Engaging students in AI-driven simulations promotes a scientific mindset, data-based reasoning, and confidence in decision-making (Kurniawan et al., 2025).

AI technology enables interactive, personalized, and data-driven learning through features such as virtual simulations, automated assessments, and adaptive instruction. In science education, AI supports critical thinking by dynamically visualizing abstract concepts (Sidek et al., 2020). Pre-service teachers with positive attitudes toward AI tend to adopt innovative, technology-based pedagogies, while those with limited digital literacy or training remain cautious in its implementation.

4.3 Main Focus of the Chapter

4.3.1 Mapping AI Literacy for K-12 Pre-service Science Teachers: Core Competencies and Training Needs in the Era of Digital Pedagogy

The global transformation of education is inseparable from the rapid advancement of digital technology and artificial intelligence (AI). Pre-service science teachers are now expected not only to master disciplinary content but also to demonstrate sufficient AI literacy (Nasaruddin, 2024). This competency is essential for implementing 21st-century pedagogies such as inquiry-based learning, complex data analysis, and the development of interactive simulations that transcend traditional teaching methods (Bustanul Arifin & Abdul Mu'id, 2024).

A strategic response to this challenge involves comprehensive and diagnostic AI literacy mapping. This process evaluates pre-service teachers' readiness and identifies gaps across cognitive, ethical, and technical dimensions of AI understanding. The findings form the basis for an in-depth needs analysis to ensure that subsequent training programs are relevant, focused, and contextually grounded (Puspitasari, 2023). As facilitators, pre-service teachers are expected to empower students to become critical consumers and responsible creators of technology (Azis, 2019).

This study adopts a sequential mixed-methods design, integrating quantitative and qualitative data to achieve a comprehensive understanding. The quantitative phase involved a survey of 55 pre-service junior high school science teachers to assess AI literacy levels and training needs. Based on these results, the qualitative phase included in-depth interviews with 10 participants and focus group discussions (FGDs) with 16–20 participants to explore perceptions, challenges, and suggestions.

Participants were purposively selected from the population of pre-service junior high school science teachers. Data were collected using an AI literacy questionnaire (five-point Likert scale) covering four competency dimensions, a training needs questionnaire, and interview and FGD protocols. Quantitative data were analyzed using descriptive and inferential statistics, while qualitative data were examined through thematic analysis to construct a holistic interpretation of AI literacy competencies and training needs.

4.4 Core AI Literacy Competencies Required by Pre-service Science Teachers

The advancement of artificial intelligence (AI) requires literacy that integrates conceptual understanding, technological proficiency, and ethical awareness. Data on pre-service science teachers' AI literacy, presented in Table 4.1, indicate an overall mean score of 3.33, categorized as good. These findings suggest that while

Table 4.1 Core AI literacy competencies required by pre-service science teachers

Competency aspects	Mean	Category
Basic AI knowledge competencies	3.36	Good
Technological skills competencies	2.89	Moderate
Pedagogical competencies (integrating AI into learning)	3.18	Good
Digital ethics competencies	3.91	Good
Average	3.33	Good

participants possess a solid conceptual and ethical foundation, their technological competencies remain underdeveloped, limiting effective classroom implementation.

Technological skills were rated at a fair level, revealing a gap between theoretical comprehension and practical application. In contrast, competencies in basic AI knowledge, pedagogical integration, and digital ethics were in the good category, reflecting growing pedagogical awareness of AI's educational potential and an understanding of ethical considerations in its use.

The primary challenge lies in operating AI-based tools such as analytical software, simulations, and educational chatbots. This highlights the necessity of capacity-building programs emphasizing hands-on experience and project-based learning to strengthen AI integration skills (Nasution, 2025).

Interview data further reveal that pre-service teachers increasingly perceive AI as an "instructional assistant" and a resource for personalized, exploratory learning rather than merely a tool for information retrieval. This perception marks a paradigm shift toward positioning AI as a collaborative partner in science education. Consequently, professional development for pre-service teachers must address not only technical proficiency but also conceptual and ethical dimensions, enabling the critical, reflective, and responsible application of AI in 21st-century science learning (Sugiarto & Farid, 2023).

4.5 AI Literacy Level of Pre-service Science Teachers

AI literacy reflects the extent to which pre-service science teachers understand, master, and apply AI concepts and technologies in educational contexts. Assessing this literacy is essential for determining their readiness to navigate digital transformation in science education, which increasingly relies on data-driven and adaptive technologies. The mapping results in Table 4.2 present AI literacy competency achievements across four indicators: basic AI knowledge, technological skills, pedagogical competence, and digital ethics and social awareness. Quantitative analysis shows an overall average score of 60.5, placing AI literacy at a **moderate level.**

For basic AI knowledge, scores fall within the moderate range, suggesting that pre-service teachers primarily perceive AI as a functional automation tool rather than a system involving complex decision-making and data analysis. This indicates the

Table 4.2 AI literacy levels of pre-service science teachers

AI literacy competency indicators	Indicator description	Average	Category
Basic AI knowledge	Understanding the concepts, working principles, and potential applications of AI in science learning	65.4	Moderate
Technological skills	Ability to use AI applications (simulations, data analysis, adaptive media) to support learning	62.8	Moderate
Pedagogical competencies	Ability to design and integrate AI into science learning scenarios	58.6	Fairly low
Digital ethics and social awareness	Understanding data privacy, algorithm fairness, and academic responsibility in the use of AI	55.2	Low
Average		60.5	Moderate

need to strengthen conceptual, pedagogical, and ethical capacities to enable reflective and strategic AI use in science learning (Fradana & Suwarta, 2025).

Technological skills were also rated moderate, with teachers' experience largely limited to basic AI applications such as ChatGPT and Canva for text and visual design. Pedagogical competence in AI integration was slightly lower, categorized as moderately low, implying that AI utilization has not yet extended to adaptive or personalized learning design. These results reveal a persistent gap between basic technical proficiency and the advanced digital competencies required for AI-enhanced science education (Aini et al., 2025).

The digital ethics and social awareness dimension scored low, indicating limited understanding of AI-related ethical issues, including data privacy, information security, and algorithmic fairness. Strengthening ethical awareness is thus essential to ensure the safe, equitable, and responsible use of AI in science learning.

4.6 The Need for Training to Improve AI Literacy Among Pre-service Science Teachers

Interview findings revealed a strong awareness among pre-service science teachers of the importance of AI literacy; however, a notable gap persists between conceptual understanding and practical application. Many participants emphasized the need for foundational AI training, as a number of peers still perceive AI primarily as chatbots rather than as tools capable of analyzing experimental data, simulating scientific processes, and providing automated feedback. This perception underscores the necessity of structured, contextualized training to facilitate the effective integration of AI into classroom practice.

The interviews also highlighted the importance of strengthening pedagogical and ethical dimensions of AI literacy. Participants expressed concern over issues such as plagiarism, data privacy, and regulatory compliance in AI use within educational settings. Ethical awareness was viewed as essential for promoting responsible and critical engagement with AI technologies and for preventing misuse among students (Beti Malia Rahma Hidayati, 2024).

Targeted professional development programs are therefore needed to enhance not only technical proficiency in AI tools but also pedagogical competence and ethical sensitivity (Astuti et al., 2025). Such training would equip pre-service science teachers to use AI reflectively, innovatively, and responsibly in supporting 21st-century science education.

4.7 Designing an AI Training Program Tailored to Science Learning in Secondary Schools

The AI literacy training program for pre-service science teachers was developed in a modular format to strengthen conceptual, technical, and ethical competencies. Implemented through a blended learning model that combined face-to-face and online sessions, the program achieved an overall mean score of 4.00, categorized as *very good*, indicating strong quality in both substance and contextual relevance. The program comprised four core modules: (1) introduction to AI in science learning, (2) practical integration of AI, (3) ethical review, and (4) evaluation of learning outcomes (see Table 4.3).

The introduction to AI module received the highest evaluation, reflecting its clarity and relevance to science education. However, the practical integration module revealed a need for further enhancement of application skills, particularly in implementing AI-based simulations and data analysis. While the blended learning approach was effective in supporting conceptual understanding, participants indicated that more extensive hands-on training would strengthen practical competence.

The program demonstrated high alignment with the secondary science curriculum, though additional support is needed for developing AI-based lesson plans, suggesting the importance of sustained mentoring. Flexible, context-sensitive training models were found to facilitate students' understanding of abstract scientific concepts

Table 4.3 AI literacy modular training design

Components or aspects assessed	Average	Category
Components of the Modular Training Design	4.03	Very good
Practices of AI Integration in Science Learning	3.97	Good
Contextualization of the Middle School Science Curriculum	4.01	Very good
Average	4	Very good

through interactive visualizations and simulations. Overall, this modular training framework shows promise for developing pre-service teachers' capacity to design innovative, reflective, and ethically grounded AI-supported science learning.

4.8 AI-Based Scenario Simulation for Pre-service Teacher Science Education: Enhancing Problem-Solving and Critical Thinking Skills in Secondary Schools

Science learning in secondary schools remains predominantly theoretical, with instruction often centered on information delivery rather than inquiry and experimentation. This approach limits students' opportunities to explore, question, and engage with scientific phenomena independently, leading to passive learning and a reliance on rote memorization. Consequently, students struggle to connect abstract concepts to real-life contexts, resulting in low engagement and superficial understanding (Ardiansyah et al., 2025).

AI-based scenario simulations offer a transformative approach to addressing these challenges. Through virtual experiments, students can safely explore hazardous chemical reactions, analyze ecological systems (Rahardhian, 2022), or test physical laws under controlled digital conditions (Andiantosa, 2024). These simulations allow repeated experimentation, variable manipulation, and real-time feedback, thereby deepening conceptual understanding and promoting inquiry-based learning.

The adoption of AI-based simulations in junior high school science education in Indonesia remains limited due to inadequate technological infrastructure and access to supporting software (Yang et al., 2022). Strategic policy support, targeted teacher training, and improved access to technological resources are therefore essential to facilitate effective implementation. When integrated meaningfully, AI simulations can make science learning more contextual, interactive, and authentic, while simultaneously cultivating students' critical thinking and problem-solving abilities that are key competencies for success in the 21st century (Yilmaz & Karaoglan Yilmaz, 2023).

The research problem addresses the effectiveness of AI-based learning through scenario simulations in improving students' thinking skills. The study focuses on comparing learning outcomes between AI-based instructional approaches and conventional teaching methods, with particular attention to the influence of simulations on critical thinking, including the ability to understand, analyze, and evaluate scientific concepts. The objective is to examine the impact of AI-based scenario simulations on students' problem-solving and critical thinking abilities.

A mixed-methods approach with a quasi-experimental design was used to evaluate the effectiveness of the intervention. The participants consisted of two groups from junior high schools with similar characteristics: an experimental group receiving AI-based scenario simulation instruction and a control group receiving conventional learning, each with 46 students.

The research instruments included a case-based problem-solving test with an analytical rubric, a critical thinking test adapted from the Cornell Critical Thinking Test, a Likert-scale perception questionnaire, interview guides for teachers and students, and classroom observation sheets aligned with indicators of problem-based learning and critical thinking. Quantitative data were analyzed using independent t-tests with a significance level of 0.05, while qualitative data from interviews and observations were analyzed thematically to identify patterns in experiences, perceptions, and implementation challenges. The integration of both data types provided a comprehensive understanding of the effectiveness, acceptability, and developmental potential of AI-based scenario simulations in secondary school science education.

4.9 Effect of AI-Based Learning with Scenario Simulations on Students' Problem-Solving Abilities

The normality test results indicated that the data on problem-solving abilities were normally distributed, with significance values of 0.39 for the control group and 0.353 for the experimental group, thus fulfilling the assumptions for parametric analysis. The independent samples t-test yielded a significance value of 0.000 (<0.05), indicating a statistically significant difference between the AI-based and conventional classes in improving students' problem-solving abilities.

The average posttest score of the AI-based group was 84.67, compared to 51.91 in the conventional group, with a mean difference of 32.785. This finding demonstrates the effectiveness of AI-based learning in enhancing students' ability to solve problems comprehensively. Students in the AI-based group showed improvement in identifying problems, formulating hypotheses, and systematically evaluating solutions. The interactive nature of AI-based scenario simulations encourages in-depth analysis and data-driven decision-making. In contrast, conventional learning remains focused on procedural repetition, leading students to memorize solutions without a deep conceptual understanding or the ability to transfer knowledge to new contexts.

AI-based learning exposes students to diverse, contextualized scenarios that cultivate adaptive thinking strategies. Overall, the findings show that AI-based learning significantly enhances problem-solving skills through reflective, analytical, and creative thinking processes. Students not only master problem-solving procedures but also develop the capacity to address complex situations adaptively. When applied pedagogically, AI technology serves as a strategic tool for fostering sustainable higher-order thinking skills (Yang et al., 2022).

Moreover, AI-based scenario simulations provide dynamic and interactive learning experiences aligned with the principles of problem-based learning, positioning students as active participants in constructing knowledge. The interaction between students and the AI system promotes reflective, collaborative, and evidence-based learning (Amrullah et al., 2025). Consequently, students transition from passive recipients of information to active discoverers of meaning through exploration and problem-solving processes (Kim et al., 2022).

4.10 Effect of AI-Based Learning with Scenario Simulations on Students' Critical Thinking Skills?

The normality test results showed that the data on critical thinking skills were normally distributed, with significance values of 0.180 for the control group and 0.616 for the experimental group, meeting the assumptions for parametric analysis. The Independent Samples t-test produced a 2-tailed significance value of 0.000 (<0.05), indicating a significant difference between the AI-based and conventional classes. These results demonstrate that AI-based learning has a substantial positive effect on the development of students' critical thinking skills.

Students engaged in AI-based scenario simulations demonstrated greater proficiency in analyzing information, identifying causal relationships, and evaluating the credibility of data sources. Simulation-based discussions encouraged them to question assumptions, compare alternative solutions, and assess the validity of arguments that arose during learning activities. Consequently, students developed not only conceptual understanding but also the ability to evaluate, draw conclusions, and refine their own reasoning processes (Nasution & Mujib, 2022).

AI-based learning fosters active participation and engagement in critical inquiry. Through hands-on experience, students learn to test ideas and draw conclusions grounded in evidence rather than merely receiving information (Kurniawan et al., 2025). Overall, the findings confirm that AI-based learning significantly enhances students' critical thinking abilities. The integration of AI technology in education represents not only a technological advancement but also a pedagogical transformation that strengthens students' higher-order thinking skills (Indrawan & Marvida, 2023).

4.11 Pre-service Science Teachers' Attitudes and Practices toward AI Integration in Secondary Education: A Mixed-Methods Study

AI has the potential to enhance the quality of science education through experimental simulations, automated data analysis, and adaptive learning systems that adjust to individual student needs (Kotsis, 2025). This technology promotes interactive, personalized, and data-driven learning experiences while fostering creativity and scientific reasoning. Beyond serving as an assistive tool, AI functions as a learning partner that supports pre-service teachers in designing inquiry-based instruction and providing immediate feedback on student performance (Lampou, 2023).

The success of AI integration in educational contexts depends largely on teachers' attitudes and classroom practices (Tabiin, 2021). Pre-service teachers must possess sufficient understanding, readiness, and positive dispositions toward technology to utilize it effectively in lesson planning, instruction, and assessment (Akinwalere & Ivanov, 2022). Accordingly, this study aims to examine the attitudes and practices of pre-service science teachers regarding AI implementation in secondary education, thereby providing empirical insights to inform educational policy, teacher training, and strategies for effective, ethical, and sustainable AI integration.

A mixed-methods design with a sequential explanatory approach was employed to obtain a comprehensive understanding of teachers' attitudes and practices. The quantitative phase involved a survey of 60 pre-service secondary science teachers selected through stratified random sampling to capture variations by region and school type. The findings from this phase informed the qualitative phase, which included in-depth interviews and focus group discussions (FGDs) with 10 pre-service teachers selected purposively based on their experience with AI implementation. Research instruments consisted of a Likert-scale questionnaire measuring perceived usefulness, ease of use, ethics, and trust in technology, along with interview and FGD protocols exploring experiences, challenges, and recommendations for AI integration in science learning.

4.12 Pre-service Science Teachers' Attitudes Toward Integrating AI into Learning

Attitude represents a key indicator of readiness for technology integration in educational contexts. A positive attitude reflects awareness of AI's potential to enhance the effectiveness and quality of science learning (Gligorea et al., 2023), whereas hesitation often indicates limited understanding or experience in its application. In this study, attitudes were assessed across several dimensions: perceived usefulness, ease of use, ethical considerations, belief in AI's role in education, and openness to technological innovation.

The results showed that the mean score for perceived usefulness was 3.39 (good), indicating that pre-service teachers generally recognize the benefits of AI in improving learning effectiveness. Ease of use and readiness were categorized as fair, suggesting that some participants still feel unprepared to apply AI consistently in classroom practice due to limited technical competence. Ethical attitudes were rated as good, reflecting awareness of issues such as data privacy, fairness, and academic integrity. Meanwhile, belief in AI's educational role and openness to technological innovation were both rated as very good. Variations in attitudes were associated with factors such as age, teaching experience, and digital literacy (K. Kotsis, 2024).

Teachers with greater technological familiarity demonstrated higher openness to innovation, while those with limited experience tended to be more cautious. Digital readiness emerged as a significant determinant of AI acceptance (Suariqi Diantama, 2023). To transform positive attitudes into practical competence, continuous training and mentoring are essential to deepen understanding of AI's pedagogical applications and to develop technical proficiency. A sustained positive attitude can ultimately support innovative, effective, and ethical science instruction.

Interview data reinforced these findings, revealing that most pre-service science teachers hold favorable views toward AI integration, recognizing its potential to facilitate learning, provide automatic feedback, and enrich instruction through interactive simulations. However, a minority expressed reservations due to limited technical expertise, ethical concerns, and apprehension about AI's potential to diminish the teacher's role.

4.13 Actual Practices of Pre-service Science Teachers Using AI in the Classroom

Actual classroom practices reflect the extent to which AI technology has been effectively implemented in instructional activities. Measurements were conducted to assess the frequency and forms of AI utilization, including the development of teaching materials, use of experimental simulations, application of automated assessments, and engagement with AI as a learning assistant. Based on interview data and quantitative analysis, the overall implementation of AI among pre-service science teachers was categorized as fairly good, with an average score of 3.08 across all aspects.

The use of AI for searching and organizing teaching materials received the highest mean score, categorized as good, indicating that pre-service teachers are relatively active in employing technology to support lesson preparation. Meanwhile, the use of AI in experimental simulations, concept visualization, automated assessment, learning outcome analysis, lesson plan design, and as a learning assistant were rated as fair. These results suggest that AI integration in instructional practice remains inconsistent and requires more systematic application.

Several factors constrain the implementation of AI-based practices, including limited infrastructure, insufficient training, and lack of institutional policy support (Alam et al., 2024). Many pre-service teachers reported inadequate access to technological equipment and stable internet connections, as well as limited exposure to structured training on AI integration in science education (Akinwalere & Ivanov, 2022). Consequently, AI use remains largely a personal initiative, primarily undertaken by individuals with higher levels of digital literacy (OECD, 2023).

In practice, pre-service teachers predominantly utilize accessible tools such as ChatGPT, Google Gemini, and PhET Simulations to assist in material preparation, question development, and virtual experimentation. Although these practices demonstrate growing interest and awareness of AI's potential, their application remains focused on practical and supportive functions rather than full integration into pedagogical design, curriculum development, or data-driven assessment processes.

4.14 Relationship Between Pre-service Science Teachers' Attitudes and Practices in Integrating AI

An analysis was conducted to examine the relationship between pre-service science teachers' attitudes toward AI and their actual practices in implementing AI within learning activities. The results indicated a moderate positive correlation ($r = 0.68$), suggesting that more favorable attitudes toward AI are associated with higher levels of practical application.

The results indicate that more positive pre-service teachers' attitudes are toward the benefits and potential of AI are followed by greater likelihood of them implementing it in instructional activities. Teachers who hold favorable views of AI tend to be more proactive in utilizing applications such as ChatGPT, Google Gemini, and PhET Simulations to explain scientific concepts, develop assessment items, and enrich interactive learning media (Lund & Wang, 2023). A positive attitude thus serves as a key prerequisite for the successful integration of AI in classroom practice.

However, the correlation between attitudes and practices is not entirely strong, suggesting that AI use is also shaped by external factors such as inadequate infrastructure, unstable internet access, limited technical training, and insufficient policy support from educational institutions. Strengthening AI implementation therefore requires a comprehensive approach that combines the cultivation of positive attitudes with the enhancement of digital competence and institutional support (K. Kotsis, 2024). Adaptive school policies, adequate infrastructure, and continuous professional development are essential to build teachers' confidence and ability to utilize AI effectively (Lund & Wang, 2023). These efforts not only reinforce the connection between attitudes and practices but also foster the development of innovative, efficient, and future-oriented science learning aligned with the demands of 21st-century education (Sidek et al., 2020).

4.15 Solutions and Recommendations

Future initiatives to enhance AI literacy among pre-service K-12 science teachers should prioritize the development of practical implementation skills. This can be achieved by strengthening contextual, hands-on training and integrating the design and application of AI-based learning plans to bridge the gap between conceptual knowledge and classroom practice. A blended learning approach should serve as the primary framework that ensures the alignment between training materials with the curriculum standards and pedagogical goals.

The use of AI-driven learning through scenario-based simulations has demonstrated effectiveness in enhancing students' problem-solving and critical thinking abilities, indicating its potential for broader integration into secondary science education. Successful implementation, however, requires concurrent improvements in digital literacy and pedagogical competence among pre-service teachers to ensure AI is applied effectively and ethically. Institutional and governmental support is essential in providing infrastructure, professional development, and regulatory policies that promote the adoption of adaptive learning technologies.

Moreover, fostering positive attitudes toward AI necessitates continuous, practice-oriented literacy programs alongside the dissemination of ethical guidelines that safeguard privacy and responsible use. To translate positive perceptions into effective classroom practice, stakeholders must ensure access to basic digital infrastructure (devices and internet connectivity), offer structured guidance on incorporating AI into lesson plans, and deliver applied training that emphasizes tangible, classroom-relevant applications.

4.16 Future Research Directions

Future studies should focus on developing specialized AI literacy training modules that emphasize practical classroom integration and the creation of AI-based lesson plans. Evaluating the effectiveness of post-training mentoring models is also essential to ensure the sustainability of AI implementation in teaching practice. Moreover, future research should examine the extent to which such training improves student competencies and enhances overall learning quality.

The integration of AI in science education highlights the need to strengthen technological literacy and pedagogical competence among pre-service K-12 teachers. These findings provide a foundation for policymakers and educational institutions to advance AI-driven innovations that foster critical thinking and problem-solving skills. Subsequent research should broaden the scope of inquiry by exploring diverse AI-based simulation models and their effects on affective and collaborative learning dimensions to gain a more holistic understanding of AI's pedagogical impact.

Further investigation is also warranted into the effectiveness of digital literacy and pedagogical mentoring programs that incorporate AI tools. In addition, future research should examine contextual factors; such as institutional policies, technological infrastructure, and digital culture that shape the successful implementation of AI in classrooms. Insights from such studies can inform the formulation of sustainable, evidence-based policies and learning models that effectively integrate AI into science education.

4.17 Conclusion

The AI literacy level of pre-service K-12 science teachers is categorized as moderate, indicating that their understanding remains at an introductory stage and is not yet fully integrated into teaching practice. The implementation of a modular blended learning–based training program has proven effective in enhancing AI literacy, although a clear gap persists between conceptual understanding and practical application, particularly in AI integration, hands-on practice, and lesson plan development.

AI-supported scenario-based simulation learning has been shown to improve students' problem-solving and critical thinking skills in secondary science education. Such simulations create interactive and contextual learning environments that encourage the application of scientific concepts in authentic situations. Statistical analyses confirm that AI-based learning produces significantly better outcomes than conventional methods, strengthening higher-order thinking skills, learning independence, and motivation.

Pre-service science teachers generally demonstrate positive attitudes toward the use of AI in education and recognize its potential to support personalized learning, experimental simulations, and efficient analysis of learning outcomes. Despite this openness, AI application in classroom practice remains limited and has not yet been systematically integrated into lesson planning or instruction. The findings indicate a positive correlation between teachers' attitudes and their actual use of AI, with those holding favorable views showing greater engagement in AI-based learning activities. However, effective implementation is still constrained by external factors such as limited infrastructure, insufficient training, and inadequate institutional and policy support.

References

Aini, M., Pabur, H. E., & Hibrida, A. R. (2025). Sosialisasi pemanfaatan ai bagi guru sma negeri 2 tomohon: upaya optimalisasi pembelajaran berbasis teknologi. *Jurnal Pengabdian Kepada Masyarakat, 4*(3), 133–142.

Akinwalere, S. N., & Ivanov, V. (2022). Artificial intelligence in higher education: challenges and opportunities. *Border Crossing*. https://doi.org/10.33182/bc.v12i1.2015

Alam, S., Hameed, A., Madej, M., & Kobylarek, A. (2024). Perception and practice of using artificial intelligence in education: An opinion based study. *XLinguae*. https://doi.org/10.18355/XL.2024.17.01.15

Amrullah, J. D. R., Yusmar, F., Astuti, S. R. D., Ahmad, N., & Putra, P. D. A. (2025). Pelatihan pbl (project based learning) berbantuan vr (virtual reality) berbasis ai untuk guru ipa di kabupaten jember. *Bima Abdi: Jurnal Pengabdian Masyarakat, 5*(3), 564–572. https://doi.org/10.53299/bajpm.v5i3.2333

Andiantosa, S. W. (2024). Pengembangan e-lkpd terintegrasi video eksperimen dan kegiatan laboratorium virtual untuk melatihkan keterampilan proses sains siswa pada topik gerak harmonik sederhana. *Multi Discere Journal, 3*(2), 4.

Ardiansyah, R., Suprapto, P. K., Fitriani, R., Diella, D., Suhendi, Y., Siliwangi, U., Fisika, P., & Siliwangi, U. (2025). *Integrasi artificial intelligence sebagai inovasi perancangan perangkat pembelajaran dan asesmen guru. 6*(1), 61–69. https://doi.org/10.29408/ab.v6i1.27964

Astuti, A., Thoha, M., Dahliah, J., Maryanti, A., Ambarita, D., Rifa'i, R., & Hidayat, T. (2025). Etika penggunaan ai di sekolah: menyeimbangkan inovasi dengan integritas akademik. *RIGGS: Journal of Artificial Intelligence and Digital Business, 4*(2), 5893–5900. https://doi.org/10.31004/riggs.v4i2.1639

Azis, T. N. (2019). Strategi pembelajaran era digital. *Annual Conference on Islamic Education and Social Sains (ACIEDSS 2019), 1*(2), 308–318.

Beti Malia Rahma Hidayati, I. N. S. (2024). Etika pemanfaatan artificial intelligence dalam pendidikan: mendidik generasi yang bertanggung jawab terhadap teknologi. *Jurnal Ilmu Sosial Sains Dan Teknologi, 2*(2), 66–80.

Bustanul Arifin, & Abdul Mu'id. (2024). Pengembangan kurikulum berbasis keterampilan dalam menghadapi tuntutan kompetensi abad 21. *DAARUS TSAQOFAH Jurnal Pendidikan Pascasarjana Universitas Qomaruddin, 1*(2), 118–128. https://doi.org/10.62740/jppuqg.v1i2.23

Fradana, A. N., & Suwarta, N. (2025). Artificial intelligence driven literacy practices in early language education. *Academia Open, 10*(1), 1–16. https://doi.org/10.21070/acopen.10.2025.11438

Gligorea, I., Cioca, M., Oancea, R., Gorski, A. T., Gorski, H., & Tudorache, P. (2023). Adaptive learning using artificial intelligence in e-learning: A literature review. *Education Sciences*. https://doi.org/10.3390/educsci13121216

Indrawan, D., & Marvida, T. (2023). Kompetensi pedagogik guru dalam pemanfaatan model pembelajaran simulasi berbasis tik. *Educative: Jurnal Ilmiah Pendidikan*. https://doi.org/10.37985/educative.v1i1.6

Kim, J., Lee, H., & Cho, Y. H. (2022). Learning design to support student-AI collaboration: Perspectives of leading teachers for AI in education. *Education and Information Technologies*. https://doi.org/10.1007/s10639-021-10831-6

Koh, J., Cowling, M., Jha, M., & Sim, K. N. (2023). The human teacher, the ai teacher and the aied-teacher relationship. *Journal of Higher Education Theory and Practice*. https://doi.org/10.33423/jhetp.v23i17.6543

Kotsis, K. (2024). Integrating artificial intelligence in science education: Benefits and challenges. *International Journal of Educational Innovation, 6*(3), 39–49. https://doi.org/10.69685/icas1772

Kotsis, K. T. (2025). Integrating artificial intelligence for science teaching in high school. *LatIA, 3*. https://doi.org/10.62486/latia202589

Kurniawan, D., Masitoh, S., Bachri, B. S., Warman, Kamila, V. Z., Subastian, E., & Wahyuningsih, T. (2025). Integrating AI in digital project-based blended learning to enhance critical thinking and problem-solving skills. *Multidisciplinary Science Journal, 7*(12). https://doi.org/10.31893/multiscience.2025552

Lampou, R. (2023). The Integration of artificial intelligence in education: Opportunities and challenges. *Review of Artificial Intelligence in Education*. https://doi.org/10.37497/rev.artif.intell.educ.v4i00.15

Lund, B. D., & Wang, T. (2023). Chatting about ChatGPT: How may AI and GPT impact academia and libraries? In *Library Hi Tech News*. https://doi.org/10.1108/LHTN-01-2023-0009

Nasaruddin, M. (2024). Peran kecerdasan buatan terhadap transformasi pendidikan di era digital. *SALIMIYA: Jurnal Studi Ilmu Keagamaan Islam, 5*(4), 410–418.

Nasution, S. R., & Mujib, A. (2022). Peningkatan kemampuan pemecahan masalah matematis dan kemandirian belajar siswa melalui pembelajaran berbasis masalah. *Edumaspul: Jurnal Pendidikan*. https://doi.org/10.33487/edumaspul.v6i1.1850

Nasution, W. R. (2025). Integrasi mata pelajaran coding dan kecerdasan buatan (ai) dalam kurikulum sekolah dasar sebagai upaya meningkatkan keterampilan abad ke-21 wilda. *Journal of Community Dedication, 4*(4), 225–236.

OECD. (2023). *PISA 2022 results (volume I): The state of learning and equity in education* (Vol. 1). In *OECD Publishing* (Vol. 46, Issue 183).

Puspita, V., Marcelina, S., & Melindawati, S. (2023). Pelatihan penggunaan artificial intellegence dalam penyusunan modul pembelajaran bagi guru sekolah dasar. *BHAKTI NAGORI (Jurnal Pengabdian Kepada Masyarakat)*. https://doi.org/10.36378/bhakti_nagori.v3i2.3402

Puspitasari, Y. (2023). Analisis kebutuhan pelatihan berdasarkan kompetensi pramuwisata di jawa tengah (competency-based training needs analysis of tour guide in central java). *Jithor, 6*(1), 105–120.

Rahardhian, A. (2022). Kajian kemampuan berpikir kritis (critical thinking skill) dari sudut pandang filsafat. *Jurnal Filsafat Indonesia*. https://doi.org/10.23887/jfi.v5i2.42092

Sidek, R., Halim, L., Buang, N. A., & Mohamad Arsad, N. (2020). Fostering scientific creativity in teaching and learning science in schools: A systematic review. *Jurnal Penelitian Dan Pembelajaran IPA*. https://doi.org/10.30870/jppi.v6i1.7149

Suariqi Diantama. (2023). Pemanfaatan artificial intelegent (ai) dalam dunia pendidikan. *DEWANTECH Jurnal Teknologi Pendidikan*. https://doi.org/10.61434/dewantech.v1i1.8

Sugiarto & Farid, A. (2023). Model pembelajaran sosiodrama untuk meningkatkan minat belajar siswa sd. *Jayapangus Press Cetta: Jurnal Ilmu Pendidikan, 6*(3), 580–597.

Tabiin, P. (2021). Student's perception toward the use of edlink platform at the second semster of english program at iain parepare. *Frontiers in Neuroscience*.

Yang, J., Shi, G., Zhuang, R., Wang, Y., & Huang, R. (2022). 5G and smart education: Educational reform based on intelligent technology. *Frontiers of Education in China*. https://doi.org/10.3868/s110-007-022-0021-4

Yilmaz, R., & Karaoglan Yilmaz, F. G. (2023). The effect of generative artificial intelligence (AI)-based tool use on students' computational thinking skills, programming self-efficacy and motivation. *Computers and Education: Artificial Intelligence*. https://doi.org/10.1016/j.caeai.2023.100147

Chapter 5
Design Thinking in Science Education: Preparing Pre-Service Teachers for Human-Centered Innovation in the Age of Artificial Intelligence

Funda Ergulec and Rahime Filiz Agmaz

5.1 Introduction

The advanced use of artificial intelligence (AI) technologies has reshaped the meaning of education in all subjects, including science education. While digital transformation has traditionally involved the adoption of networked systems, digital content, and data management tools, the emergence of AI marks a shift toward technologies that do not merely store or deliver information but actively generate, interpret, and reorganize it (Haenlein & Kaplan, 2019). This transition from "digitization" to "computational intelligence" is crucial for understanding why AI requires new pedagogical, ethical, and organizational approaches.

Artificial intelligence can be defined as computational systems capable of performing tasks that typically require human cognitive processes, such as perception, pattern recognition, data-driven inference, and adaptive decision-making (Haenlein & Kaplan, 2019). In educational contexts, AI's distinctiveness lies in its capacity to process large volumes of learner-generated data, model complex systems, and generate predictive insights that support instructional and cognitive tasks in ways that conventional digital tools cannot (Holmes et al., 2022; Luckin et al., 2022).AI incorporates algorithmic reasoning and autonomous optimization, enabling forms of augmentation that extend beyond simple tool use.

AI is distinguished from traditional educational technology by its unique effectiveness in the processing of data, modeling complex systems, and predicting outcomes.

F. Ergulec (✉)
Eskisehir Osmangazi University, Eskisehir, Turkey
e-mail: fundaergulec@gmail.com

R. F. Agmaz
Necmettin Erbakan University, Konya, Turkey

© The Author(s), under exclusive license to Springer Nature Switzerland AG 2026

M. Alkış Küçükaydın (ed.), *Preparing K-12 Pre-Service Science Teachers for the Age of Artificial Intelligence*, Advances in Technology-Rich Science Education 4,
https://doi.org/10.1007/978-3-032-18030-8_5

After the inclusion of AI technologies in education, the challenges of teaching science in K-12 schools are not simply a matter of teaching students' scientific content but also of preparing students to exist in a world where students' creativity is augmented by computational intelligence (Holmes et al., 2022; Luckin, 2020). Therefore, new pedagogical tools are needed to foster creativity, problem-solving, and ethical awareness in pre-service science education and teachers who will educate future students. DT is an approach that has emerged as a human-centered methodology complementary to scientific research (Henriksen et al., 2020; Razzouk & Shute, 2012). DT applications in education have the potential to improve problem-solving skills of students and develop their capabilities to provide more innovative and creative solutions to real-world and unusual challenges (Brenner et al., 2016; McLaughlan & Lodge, 2018).

DT, in addition to its methodological appeal, also serves as a conceptual bridge between scientific reasoning and socio-technological innovation. In science education, students frame complex problems from multiple perspectives, generate and prototype new ideas, and evaluate these ideas through experiments in a repetitive manner (Henriksen et al., 2017). When integrated into pre-service teacher education, DT has the potential to enable future science teachers to engage their students in not only what to learn but also how to think, developing agency, empathy, and reflective decision-making in technology-mediated learning environments (Carroll et al., 2010; Henriksen et al., 2017; Koh et al., 2015). In an education system where AI has been widely used, DT can serve as a pedagogical framework to help pre-service science teachers deal with the ethical, creative, and cognitive dimensions of AI-enhanced education (Chai, 2021; Holmes et al., 2022).

Thus, this chapter explores how DT can be used in pre-service science teacher education in ways that are responsive to the demands of the AI era. The inclusion of DT in science teacher education can offer a platform that fosters ethical consideration of the uses of AI in scientific and social contexts while creating creative and critical thinking. By combining DT's organized creativity with the rigor of scientific thought and the promise of AI, teacher educators can prepare future science teachers to assist in innovative, ethical, and human-centered practices in their classrooms (Henriksen et al., 2020). In doing so, DT becomes not only a pedagogical tool but a way of thinking that positions teachers and students as co-designers of meaningful learning experiences in the age of AI.

5.2 Phases of Design Thinking

Plattner (2010) describes the process of DT in terms of five iterative stages; Empathize, Define, Ideate, Prototype, and Test; each of which is guided by the questions what, why, and how in a bid to promote better understanding for designers. Users such as designers, or teachers in this case, can move through the stages of the process seamlessly, going back or re-entering specific stages as needed, in a bid to better understand problems and develop potential solutions.

5.3 Empathize: Human and Context-Centered Understanding

The empathize phase of DT concentrates on users' experiences, perspectives, and emotional or practical needs (Plattner, 2010). According to Plattner (2010), empathy involves three interrelated practices: observation, engagement, and listening which include observing users' actions and contexts, engaging through interaction to build contextual understanding, and synthesizing these insights through reflective listening.

5.4 Define: Reframing the Problem

The Define stage requires learners to synthesize insights from the empathy phase and frame complex, ambiguous problems in new ways. This stage involves creating a problem statement and a persona, a fictional but evidence-based representation of a user to clarify needs and goals. Designers often revisit this stage to refine problem statements as their understanding deepens.

5.5 Ideate: Generating Creative Ideas

The Ideate phase focuses on developing a wide range of possible solutions through creative exploration. Techniques such as brainstorming, storyboarding, and concept mapping help designers move beyond conventional thinking (Plattner, 2010). The key is not to find a single "correct" answer but to encourage collaborative divergence, using imagination, interdisciplinary perspectives, and unexpected combinations to generate breakthrough ideas.

5.6 Prototype: Making Ideas Tangible

Following idea generation, designers implement their ideas in the Prototype mode by modeling a potential solution or solutions to the problem. It is a chance to explore giving concepts a tangible form rather than an effort to get a definitive conclusion. Prototyping allows designers to translate abstract ideas into tangible forms, whether through role-playing, storyboards, digital materials, or AI-based simulations. Plattner (2010) identifies prototyping as a reflective process, one that enables designers to think, communicate, and problem-solve through making. Rather than aiming for perfection, prototypes should rapidly test key ideas and invite feedback.

5.7 Test: Evaluation and Refinement

The Testing phase, which constitutes the fifth mode of the design process, involves evaluating the prototype with actual end-users or relevant stakeholders. The final Test stage involves gathering feedback from users to refine and improve the prototype. Testing is also a renewed opportunity for empathy, as designers observe how users interact with their creations, uncovering hidden needs and misalignments (Plattner, 2010). Designers employ various methods during this stage, such as interviewing users, observing their direct interaction with the prototype, or utilizing other techniques to gather constructive feedback necessary for refining the proposed solution(s). This crucial testing may reveal that the designer may need to significantly revise the prototype, or perhaps even revisit and redefine the initial problem statement or point of view. Because design is inherently iterative, feedback received at this stage may prompt a return to the empathize mode to gain deeper user understanding, or a shift back to the ideate mode to explore alternative concepts. At any point, a designer is free to repeat or reconsider a preceding phase based on the insights gained. The goal is not merely to assess whether the prototype "works," but to understand *how* it can be improved and *why* certain design decisions succeed or fail (Plattner, 2010).

5.8 The Role of Design Thinking in Science Education

DT is both a structured and adaptable approach that helps address complex or ill-defined problems, those without a single correct answer but that demand exploration, creativity, and iteration (Dorst, 2011; Razzouk & Shute, 2012). The commonly used DT cycle described earlier parallels the process of scientific inquiry, such as observing, formulating hypotheses, experimenting, and evaluating, but goes further by emphasizing empathy, imagination, and repeated refinement (Henriksen et al., 2020; Scheer et al., 2012).

In science education, the use of DT offers several pedagogical benefits. It encourages students to frame scientific issues in socially meaningful and ethical ways, helping them perceive science not only as an abstract discipline but also as a means of addressing human and environmental challenges (Crismond & Adams, 2012; Vongkulluksn et al., 2018). It also fosters collaboration and divergent thinking, prompting learners to examine multiple possibilities and build resilience through collective creativity during experimentation (Henriksen et al., 2020). Moreover, DT nurtures reflection and metacognition as students move back and forth between generating ideas, prototyping, and testing, gradually learning to interpret failure as a valuable part of the discovery process (Kolodner, 2002).

While scientific inquiry seeks to uncover generalizable truths about the natural world, DT starts from a human-centered stance, situating scientific understanding within real-world and social contexts (Carroll et al., 2010). Scientific inquiry often highlights hypothesis testing and evidence-based reasoning, whereas DT draws attention to the human and societal dimensions of solving problems. When the two are integrated, pre-service science teachers can design learning experiences that help students not only comprehend scientific ideas but also apply them thoughtfully in contexts marked by uncertainty, complexity, and ethical concerns. In this sense, DT may serve as a bridge between analytical scientific thinking and creative innovation, preparing future educators to cultivate scientific literacy, empathy, and adaptability in an increasingly AI-enhanced world.

5.9 Design Thinking in the Age of Artificial Intelligence

Combining DT and AI can offer a strong, groundbreaking approach to science education today, specifically to pre-service teacher education. Both enable experimentation, iterative refinement, and solving complex problems but from different yet complementary perspectives. Where AI is particularly effective in processing large amounts of data, modeling complex systems, and predicting outcomes, providing essential data skills for scientific citizenship (Holmes et al., 2022; Luckin, 2020), DT keeps innovation connected with human experience, making sure technology stays with empathy and ethical awareness (Holmes et al., 2022; Luckin, 2020). Together, they can promote human-centric innovation in all areas of design cycle:

- Data-Enhanced Empathy: Students are supported by AI technology to model complex science and social problems (e.g., climate change) using interactive simulations and forecast models (Holmes et al., 2022). When used in the DT empathy phase, this data skill can help to find real user needs and link science ideas to real-life human and environment situations, and support learners to develop scientific citizenship.
- AI-Driven Ideation and Prototyping: Generative AI models act as cognitive partners in divergent thinking, accelerating visualization of instructional design or experimental procedures (Luckin, 2020; Zawacki-Richter et al., 2019).
- Ethical and Reflective Practice: This intersection requires critical consideration of the ethical and social implications of technology. AI technology provokes important questions regarding equity, bias, and accountability (Tuomi et al., 2023; Williamson & Eynon, 2020). Embedding such ethical questioning within the design process may develop technological fluency and moral sensitivity in pre-service teachers.

5.10 Lesson Design: Design Thinking for Pre-service Teachers in the Age of AI

The following lesson plan illustrates the integration of DT into pre-service teacher education to enable innovation, creativity, and ethical problem-solving in the AI age. It introduces a step-by-step process that leads pre-service teachers through the iterative cycle of empathizing with students, framing authentic problems from the real world, developing and prototyping solutions, and assessing solutions in real or simulated classrooms. DT framework in this lesson plan, not only emphasizes scientific inquiry but also cultivates human-centered thinking, collaboration, and adaptive problem-solving skills essential for preparing future educators to navigate and shape AI-enhanced learning environments.

The course mentioned here, Design Thinking course, was part of the undergraduate teacher education program in a public university in Türkiye. The course was designed and delivered by the first author under a broader initiative to integrate DT in teacher education. The course was organized into five general phases following the Stanford Design Thinking process, Empathize, Define, Ideate, Prototype, and Test (Plattner, 2010), with additional introductory and reflection components being inserted within these phases. Each phase took approximately about two weeks, involving class discussion, group work, and feedback. Every step was supported by carefully selected readings bridging DT principles. Pre-service teachers engaged in whole-class and small-group discussions, analyzing how design processes could be applied to challenges such as laboratory safety, conceptual misunderstandings, laboratory design, or equitable participation in STEM learning.

The course included a final project where pre-service teachers worked in small teams to address real problems from science classrooms, following the iterative stages of DT. Throughout this semester-long project, each group of 3–4 pre-service science teachers selected a problem of practice from its teaching environment, for instance, a challenge observed during teaching practice, laboratory lessons, or science learning procedures. Throughout the semester, students progressively worked on the problem using the steps of the Stanford Design Thinking model (empathize, define, ideate, prototype, test, and reflect) (Plattner, 2010). Each phase had specific deliverables demonstrating their work in the process, including an empathy report of observations or interviews for the Empathize phase, a problem definition statement for the Define phase, a record of brainstorming with peers or mentors for the Ideation phase, and a low-fidelity prototype or classroom intervention plan for the Prototype phase. The final project aimed to help pre-service teachers apply DT to authentic educational challenges, generate creativity, empathy, and iterative problem-solving in science education.

Course Objective

The primary purpose of this course is to develop pre-service science teachers' DT capability to enable them to generate innovative, student-centered solutions to pedagogical or technology problems encountered in the teaching process.

Phase 1: Empathizing and Identifying the Problem (Weeks 1–2)

Objective:

To help pre-service teachers gain a deep understanding of a problem in the educational context and develop insights into the needs and experiences of target users (students, teachers, or parents).

Activity Flow:

- Students form groups of 3–4 members.
- Each group identifies an educational problem area (e.g., laboratory safety awareness, conceptual misconceptions in science classes, gender equality in STEM activities).
- During the empathy phase, groups conduct interviews with target users.
- Each group prepares 6–10 interview questions and conducts interviews with at least three participants.
- Interview recordings, observations, and reflection notes are compiled into a report.

Pedagogical Rationale:

This phase promotes pre-service teachers' awareness of student needs in science education and helps them "define the problem through the user's perspective." The empathy phase is directly linked to scientific inquiry, as it involves careful observation, questioning, and interpretation, skills that form the foundation of scientific curiosity (Plattner, 2010; Razzouk & Shute, 2012).

Phase 2: Define—Reframing the Problem (Weeks 3–4)

Objective:

To synthesize the data collected during the empathy phase into a user-centered problem statement.

Activity Flow:

- Groups thematically analyze the findings from interviews and observations.
- Based on these insights, they articulate a concise problem statement that reflects users' real needs and challenges.
- Each group formulates a "How might we…?" question that frames the issue as an opportunity for innovation.
  - Example: *"How might we help students reduce misconceptions in science classes?"*

Pedagogical Rationale:

The Define phase supports pre-service teachers in making sense of complex educational problems through synthesis and abstraction. By translating raw empathy data into actionable insights, they learn to connect learners' needs with instructional contexts. This process nurtures analytical thinking and aligns with evidence showing that reframing problems enhances creativity and instructional adaptability (Razzouk & Shute, 2012).

Phase 3: Ideate—Generating Creative Ideas (Weeks 5–6)

Objective:

To generate a broad range of creative ideas that respond to the user-centered problem statement developed in the previous phase.

Activity Flow:

- Groups engage in structured brainstorming and lateral thinking exercises to produce multiple potential solutions.
- Ideas are recorded visually (e.g., through sketches, sticky notes, or mind maps) to encourage open-ended exploration.
- Groups select at least three promising ideas for further development in subsequent phases.

Pedagogical Rationale:

The Ideate phase fosters divergent thinking and collaborative creativity. Pre-service teachers are encouraged to move beyond conventional teaching strategies and imagine innovative, human-centered approaches to science education. This process promotes flexibility, experimentation, and critical reflection. (Henriksen et al., 2017).

Phase 4: Prototyping(Weeks 5–6)

Objective:

To transform the selected solution idea into a tangible prototype.

Activity Flow:

- Each group develops a prototype based on their selected idea.
  - Possible prototypes: digital interface designs, instructional materials, laboratory activity plans, experimental kits, or student activity posters.
- Prototypes are visually represented (e.g., sketches, mock-ups, digital models) and accompanied by descriptive explanations.
- Teams use AI-driven simulations to predict the effectiveness of their proposed interventions, followed by reflection on the scientific, ethical, and social implications of their solutions.
- Each group presents their prototype, explaining the rationale behind its design and the specific problem it addresses.

Pedagogical Rationale:

Prototyping enables pre-service teachers to translate abstract ideas into concrete instructional materials, enhancing their modeling and design abilities. This stage parallels the hypothesis–model–test cycle of scientific inquiry and encourages iterative thinking. Through this process, pre-service teachers engage not only with creativity and innovation but also with empathy and ethical reasoning, competencies essential for leading science education in the AI era (Scheer et al., 2012).

Phase 5: Testing and Feedback (Weeks 7–8)

Objective:

To evaluate prototypes through user testing and incorporate feedback to improve the final design.

Activity Flow:

- Groups test their prototypes with at least two target users.
- During testing, target users are asked reflective questions such as:
  - "How would this solution help you?"
  - "Was it easy to use?"
  - "What would you suggest we improve?"
- User feedback is synthesized into three key insights.
- Based on this feedback, each group revises and refines their prototype to produce a final version.

Pedagogical Rationale:

The testing phase strengthens pre-service teachers' experimental thinking and critical evaluation skills. By observing the impact of their designs, they engage in data-informed improvement processes. This reflective cycle mirrors scientific validation, reinforcing the iterative nature of both science and design (Morris & Warman, 2015).

Expected Learning Outcomes

By the end of this course, pre-service science teachers will be able to:

1. Develop user-centered problem-solving skills applicable to educational contexts.
2. Integrate observation, data collection, and analysis with a design-oriented perspective in science teaching.
3. Translate creative ideas into tangible educational solutions.
4. Adopt a feedback-driven approach to continuous improvement.
5. Cultivate a collaborative and empathy-based teacher identity aligned with the values of responsible innovation in the AI age.

5.11 Example Pre-service Science Teachers' Project: Enhancing Laboratory Safety Through Design Thinking

In this section, an example project developed by a team of pre-service science teachers is presented to illustrate how the DT process was applied in practice to address a real-world educational challenge. The pre-service science teachers' project is about enhancing laboratory safety and supporting students' understanding of experimental materials. It demonstrates how the structured yet flexible stages of DT, empathize, define, ideate, prototype, and test, can guide pre-service teachers in addressing real classroom challenges.

The project focused on understanding and safely using laboratory materials during science experiments in middle school settings. The pre-service teachers employed the DT to address students' challenges with laboratory material use, integrating empathy, creativity, and iterative refinement to design a practical solution.

1. **Empathize Phase**

The project began with an effort to understand the real challenges that middle school students encounter during laboratory experiments. Initial classroom observations and semi-structured interviews with three students and one science teacher revealed several recurring issues:

- Lack of clear and consistent guidance on how to handle laboratory materials safely.
- Difficulty in distinguishing between similar-looking substances or materials.
- Limited understanding of chemical compatibility, particularly confusion around reactions such as the mixing of acids and water.
- Anxiety and hesitation caused by fear of making mistakes or violating safety rules.

Through these empathy exercises, the pre-service teachers realized students' difficulty was not a lack of enthusiasm for experiments, but uncertainty and limited previous experience with materials. The teacher participant in the interviews also emphasized the absence of engaging, student-friendly tools that could help learners independently explore and identify laboratory materials before and during experiments. These findings underscored the emotional and cognitive dimensions of students' experiences in the laboratory, highlighting that safety and confidence are entwined aspects of effective science learning.

2. **Define Phase**

Based on the empathy findings, the team formulated the following problem statement:

Problem Statement:

Middle school students struggle to understand and safely use laboratory materials during experiments. They often confuse similar-looking items, are unsure of their properties, and lack accessible guidance for safe handling.

The empathy insights were synthesized into key needs:

- Students require clear, accessible explanations of what each material is and how to use it safely.
- Visual and interactive guidance can increase students' confidence and engagement.
- Safety information must be integrated into the learning process, not presented separately as rules.

This phase clarified the project's guiding question: How might we help students independently and safely identify laboratory materials during science experiments?

3. **Ideate Phase**

Based on the problem defined, the team carried out brainstorming sessions using techniques like mind mapping and "How might we…" questions to create various possible solutions. Some of the ideas that were generated are:

1. **Material Labels:** Label all materials in the lab with color-coded labels containing concise information regarding its application and useable chemicals.
2. **Interactive Button:** Install a button on a container releasing a brief audio guide on a press.
3. **Barcode System:** Enable students to scan a barcode with a tablet or smartphone to access detailed material descriptions.
4. **Portable Experiment Kit:** Create a take-home kit in which students can test material properties in a secure, controlled environment.

The group evaluated each idea using criteria such as feasibility, accessibility, and educational impact. Through collaborative discussion, the barcode system was selected as the most practical and scalable option.

4. **Prototype Phase**

Selected Idea: A barcode-based interactive system for laboratory materials.

Prototype Description:

Each laboratory material (e.g., beaker, test tube, reagent bottle) was assigned a unique barcode (see Fig. 5.1 which was generated using AI by the group). When scanned with a mobile device or tablet, the barcode linked to a digital information sheet that included:

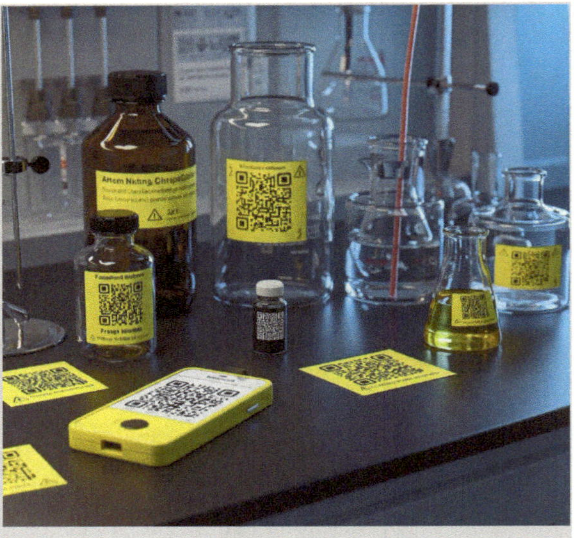

Fig. 5.1 Laboratory scene with various glass bottles and flasks on a countertop, each labeled with yellow QR code stickers. A yellow device with a large QR code is placed on the counter. The background shows laboratory equipment and a blue-tinted environment

- **Material properties** (e.g., chemical type, physical state, safety classification)
- **Usage instructions** (step-by-step guidance for experiments)
- **Safety warnings** (e.g., "Do not mix with acid," "Use gloves")
- **Compatible materials** (safe combinations for experimentation)

Rationale for Selection:

The barcode system was chosen for its simplicity, cost-effectiveness, and durability. Unlike audio or button-based solutions, it does not require specialized equipment and can easily be updated as new materials or experiments are introduced.

Needs Addressed:

- Enhances understanding of material properties.
- Reduces safety risks and confusion during experiments.
- Promotes independent learning and responsibility.
- Encourages the integration of technology in hands-on science learning.

5. **Test Phase**

The prototype was tested with two middle school students:

- **Student 1 (Grade 7, Age 12):**

  Described the design as *"practical and easy to use."* She said that scanning the barcodes before experiments can increase her confidence in handling materials. She suggested adding a transparent protective layer over the barcode to prevent it from wearing off.
- **Student 2 (Grade 6, Age 11):**

Found the design *"helpful and engaging."* He said that it can provide immediate feedback and reduce his fear of mistakes. He recommended printing the barcodes on more durable material to prevent fading over time.

Revisions and Improvements:

Based on user feedback, the team revised the prototype by:

- Using laminated barcode stickers for longevity.
- Simplifying the digital interface for faster loading.
- Adding an emergency safety tip section for high-risk materials.

Through this project, the pre-service science teacher not only addressed a tangible classroom challenge but also developed a deeper understanding of how DT can make abstract safety concepts more experiential and student-centered. The process demonstrated how empathy, iteration, and collaboration can transform a conventional instructional problem into an innovative learning opportunity. Moreover, the use of simple AI-enabled tools, such as barcode scanning and digital information retrieval, illustrates how future educators can integrate accessible technology to support safety, autonomy, and inquiry in science education.

5.12 Challenges and Opportunities

Implementing DT in pre-service teacher education presents both significant challenges and promising opportunities. On the challenge side, several common obstacles have been reported in the literature. First, time and rigid curricula traditionally limit the space for repetitive, open-ended problem-solving activities. Second, instructors and pre-service teachers may not be knowledgeable of DT methods, and in the absence of expert professional development, it can be difficult to switch from traditional teaching to designer-driven pedagogy (Eamcharoen, 2023). Within educational environments like those in Türkiye, where the system is dominated by high-stakes testing and content-focused instruction, integrating DT into the lesson plan can be more difficult to move towards pedagogic change for creativity, empathy, and non-linear problem-solving. Also, learners themselves may resist the ambiguity and open-ended nature of DT processes, particularly if they are accustomed to structured, teacher-led instruction and clear right–wrong outcomes.

However, despite these difficulties, the promise of DT is substantial and appealing. DT fosters creative confidence, collaboration, and the capacity to frame scientific problems within socially meaningful contexts (Koh et al., 2015). In the science education context, DT allows pre-service teachers to view instruction not only as the sharing of information but as human-centered problem-solving (Eamcharoen, 2023). When combined specifically with AI, understood as computational systems capable of analyzing large datasets, generating adaptive feedback, and modeling complex scientific or social processes (Holmes et al., 2022; Luckin et al., 2022), DT gains an expanded design space that traditional technologies cannot provide. AI-powered tools support pre-service teachers by allowing them to visualize scientific phenomena through dynamic simulations, explore alternative framings of problems, and receive generative suggestions during brainstorming or ideation. This AI–DT integration equips future science teachers not only with exposure to emerging technologies but also with pedagogical practices grounded in empathy, ethical reasoning, and human values. As a result, they can design learning experiences that are simultaneously cognitively rigorous, socially responsive, and technologically informed.

In the age of AI, science education extends far beyond the transmission of scientific concepts and processes. It increasingly demands that pre-service teachers cultivate a broad range of twenty-first-century competencies such as creativity, empathy, critical thinking, ethical responsibility, and adaptive problem-solving (Henriksen et al., 2017). AI intensifies these demands because its predictive modeling, pattern recognition, and generative capabilities require teachers to understand not only scientific content, but also how intelligent systems shape scientific inquiry, evidence, and uncertainty. DT offers a powerful pedagogical framework to foster these skills. Its human-centered, iterative, and problem-oriented structure enables pre-service teachers to make sense of AI's complex scientific and societal implications within meaningful educational contexts.

Over the past decade, DT has emerged as an innovative and widely adopted approach for addressing complex problems (Hennessey & Mueller, 2020; Henriksen et al., 2017). As a process that encourages empathy, creativity, innovation, collaboration, and critical thinking, DT aligns closely with key learning theories in developmental and social psychology (Albay & Eisma, 2025). Through innovative thinking, students can achieve more positive learning outcomes and develop flexible, resilient mindsets (Hennessey & Mueller, 2020). DT competencies also help learners strengthen collaboration, creativity, and adaptability in both personal and academic domains, enabling them to better cope with challenges (Thomason & Hsu, 2025).

DT places particular emphasis on problem-solving and critical thinking skills (Nguyện et al., 2025). Students learn through processes that involve experimentation, reflection, and risk-taking, understanding that failure is not an endpoint but a valuable part of the learning cycle (Vignoli et al., 2023). The iterative nature of DT involves clearly defining a problem, generating possible solutions, testing those solutions, and then refining and reevaluating them based on user feedback and reflection (Thomason & Hsu, 2025). Central to this process is the focus on developing user-centered solutions by gaining concrete insights into the problem context and user needs (Thomason & Hsu, 2025).

Empirical evidence has also demonstrated the practical relevance of DT in teacher training. Henriksen et al. (2020) described that teachers who went through DT not only acquired cognitive gains but also affective: they were more empathetic, more effective in interacting with students, and there was a shift in thinking and practice of teaching. Similarly, Lynch et al. (2021), in a study involving graduate students of entrepreneurship education, found that the use of DT combined with technological software enabled students to solve real, professionally relevant problems along with improving skills such as perspective-taking, collaboration, and adaptive problem-solving. Research has established that using AI-enabled design tasks, such as designing solutions to authentic problems, enhances empathetic understanding as well as creativity among students (Lin & Chang, 2025). Similarly, DT invites learners to move beyond surface-level assumptions and make decisions grounded in user needs and contextual realities (Khan et al., 2023).

Evidence from classroom implementations further highlights DT's impact on creativity and real-world learning. Interviews with 17 DT experts revealed that teachers observed significant growth in students' creative confidence and their ability to apply DT in authentic contexts (Rauth et al., 2010). Collectively, research shows that DT enables a more complete and realistic method of addressing real-life problems (McLaughlin et al., 2022), provides a structured but challenging way of teacher professional development (Hennessey & Mueller, 2020), enhances students' creativity and problem-solving abilities (Guaman-Quintanilla et al., 2023), and is a strong self-assessment and reflective practice instrument (Vignoli et al., 2023).

For pre-service science teachers, learning to integrate AI with DT processes not only enhances instructional design but also models how they can guide K-12 learners toward creative, ethical, and innovative uses of emerging technologies. I's unique capacity to analyze learner data, detect patterns, model complex systems, and generate novel design ideas means that it functions as a cognitive partner in DT rather than merely a digital resource (Haenlein & Kaplan, 2019). In the AI era, science education extends beyond the transmission of disciplinary knowledge to include the cultivation of creativity, empathy, adaptability, and ethical responsibility (Henriksen et al., 2017). DT can provide a pedagogical framework that enables the development of these skills through its emphasis on human-centered inquiry, iterative problem-solving, and reflective practice. Its focus on collaboration and imagination aligns closely with socio-constructivist and experiential learning principles, leaning towards learners' cognitive and affective investment in authentic problem contexts (Albay & Eisma, 2025). By pairing these elements with AI-driven modeling, generative ideation, and adaptive feedback, pre-service teachers learn to design learning experiences that mirror the real scientific and ethical problems students will confront in an AI-mediated society.

Additionally, DT's cycle of problem definition, concept development, prototyping, testing, and iteration replicates the adaptive reasoning processes central to scientific literacy (Plattner, 2010; Thomason & Hsu, 2025). While working through uncertainty, risk-taking, and learning from failure, students not only learn deeper science but also internalize resilience, teamwork, and ethical consideration habits (Vignoli et al., 2023). Thus, the interaction between AI and DT within pre-service science teacher education is a potential way towards building educators who will be capable of inspiring students to apply scientific thinking and inventive design in addressing the major problems of our time.

5.13 Conclusion and Future Directions

In conclusion, DT can offer pre-service science teachers a robust framework for navigating the complexities of teaching in the AI era. By engaging in empathy with learners, reframing authentic problems, and iteratively designing innovative solutions, future teachers can cultivate classrooms that balance scientific rigor with creativity, ethical responsibility, and human-centered impact. DT not only can improve the cognitive abilities of learners but also make them think about problems in new and creative ways, developing innovative solutions in educational settings.

Looking to the future, further research is needed on the long-term implications of integrating DT and AI in teacher education. Future studies should explore professional training programs, team projects, and inter-disciplinary activities that equip pre-service teachers with skill, attitude, and technological proficiency to work in and design learning environments responsibly. Lastly, DT prepares teachers not only to adapt to technological change but to innovate it in order to facilitate learner-centered, ethical, and socially responsible science education.

References

Albay, E. M., & Eisma, D. V. (2025). Using design thinking for developing pre-service teachers' creativity in designing teaching plans to promote interactive learning in mathematics. *Learning and Instruction, 96*, 102070. https://doi.org/10.1016/j.learninstruc.2024.102070

Brenner, W., Uebernickel, F., & Abrell, T. (2016). Design thinking as mindset, process, and toolbox: Experiences from research and teaching at the University of St. Gallen. In *Design thinking for innovation: Research and practice* (pp. 3–21). Springer International Publishing. https://doi.org/10.1007/978-3-319-26100-3_1

Carroll, M., Goldman, S., Britos, L., Koh, J., Royalty, A., & Hornstein, M. (2010). Destination, imagination and the fires within: Design thinking in a middle school classroom. *International Journal of Art & Design Education, 29*(1), 37–53. https://doi.org/10.1111/j.1476-8070.2010.01632.x

Chai, C. S. (2021). Teacher professional development for science, technology, engineering and mathematics (STEM) education: A review from the perspectives of technological pedagogical content (TPACK). *Asia-Pacific Education Researcher, 30*(1), 1–10. https://doi.org/10.1007/s40299-020-00556-6

Crismond, D. P., & Adams, R. S. (2012). The informed design teaching and learning matrix. *Journal of Engineering Education, 101*(4), 738–797. https://doi.org/10.1002/j.2168-9830.2012.tb01127.x

Dorst, K. (2011). The core of 'design thinking' and its application. *Design Studies, 32*(6), 521–532. https://doi.org/10.1016/j.destud.2011.07.006

Eamcharoen, P. (2023). Design thinking for creating educational innovations for teachers in the digital age. *Journal of Industrial Education, 23*(2), C1–C12. https://doi.org/10.55003/JIE.23203

Guaman-Quintanilla, S., Everaert, P., Chiluiza, K., & Valcke, M. (2023). Impact of design thinking in higher education: A multi-actor perspective on problem solving and creativity. *International Journal of Technology and Design Education, 33*(1), 217–240. https://doi.org/10.1007/s10798-021-09724-z

Haenlein, M., & Kaplan, A. (2019). A brief history of artificial intelligence: On the past, present, and future of artificial intelligence. *California Management Review, 61*(4), 5–14. https://doi.org/10.1177/00081256198649

Hennessey, E., & Mueller, J. (2020). Teaching and learning design thinking (DT): How do educators see DT fitting into the classroom? *Canadian Journal of Education, 43*(2), 498–521. https://www.jstor.org/stable/26954696

Henriksen, D., Gretter, S., & Richardson, C. (2020). Design thinking and the practicing teacher: Addressing problems of practice in teacher education. *Teaching Education, 31*(2), 209–229. https://doi.org/10.1080/10476210.2018.1531841

Henriksen, D., Richardson, C., & Mehta, R. (2017). Design thinking: A creative approach to educational problems of practice. *Thinking Skills and Creativity, 26*, 140–153. https://doi.org/10.1016/j.tsc.2017.10.001

Holmes, W., Bialik, M., & Fadel, C. (2022). *Artificial intelligence in education: Promises and implications for teaching and learning.* Center for Curriculum Redesign.

Khan, M. A., Kazi, A. S., & Butt, A. I. (2023). Influence of design thinking in integrated STEAM inquiry-based learning on primary school students' empathy and confidence. *Global Sociological Review, 8*(1), 321–330. https://doi.org/10.31703/gsr.2023(VIII-I).32

Koh, J. H. L., Chai, C. S., Wong, B., & Hong, H. Y. (2015). *Design thinking for education: Conceptions and applications in teaching and learning.* Springer.

Kolodner, J. L. (2002). Facilitating the learning of design practices: Lessons learned from an inquiry into science education. *Journal of Industrial Teacher Education, 39*(3), 9–40. https://files.eric.ed.gov/fulltext/EJ782298.pdf

Lin, M. Y., & Chang, Y. S. (2025). Effects of design thinking STEAM instruction on AI learning and creativity. *International Journal of Technology and Design Education*, 1–23. https://doi.org/10.1007/s10798-025-09977-y

Luckin, R. (2020). *Machine learning and human intelligence: The future of education for the 21st century.* UCL Institute of Education Press.

Luckin, R., Cukurova, M., Kent, C., & Du Boulay, B. (2022). Empowering educators to be AI-ready. Computers and Education: *Artificial Intelligence, 3*, 100076. https://doi.org/10.1016/j.caeai.2022.100076

Lynch, M., Kamovich, U., Longva, K. K., & Steinert, M. (2021). Combining technology and entrepreneurial education through design thinking: Students' reflections on the learning process. *Technological Forecasting and Social Change, 164*, 119689. https://doi.org/10.1016/j.techfore.2019.06.015

McLaughlan, R., & Lodge, J. M. (2018). Facilitating epistemic fluency through design thinking: A strategy for the broader application of studio pedagogy within higher education. *Teaching in Higher Education, 24*(1), 81–97. https://doi.org/10.1080/13562517.2018.1461621

McLaughlin, J. E., Chen, E., Lake, D., Guo, W., Skywark, E. R., Chernik, A., & Liu, T. (2022). Design thinking teaching and learning in higher education: Experiences across four universities. *PLoS ONE, 17*(3), e0265902. https://doi.org/10.1371/journal.pone.0265902

Morris, H., & Warman, G. (2015). Using design thinking in higher education. *EDUCAUSE Review (Online).* Retrieved from https://www.proquest.com/scholarly-journals/using-design-thinking-higher-education/docview/3224631211/se-2

Nguyện, L. C., Hoa, H. Q., & Hien, L. H. P. (2025). Integrating design thinking into STEM education: Enhancing problem-solving skills of high school students. *Eurasia Journal of Mathematics, Science and Technology Education, 21*(4), em2611. https://doi.org/10.29333/ejmste/16084

Plattner, H. (2010). *An introduction to design thinking process guide.* Stanford University Institute of Design (d.school).

Rauth, I., Köppen, E., Jobst, B., & Meinel, C. (2010). Design thinking: An educational model towards creative confidence. In *DS 66-2: Proceedings of the 1st International Conference on Design Creativity (ICDC 2010).*

Razzouk, R., & Shute, V. (2012). What is design thinking and why is it important? *Review of Educational Research, 82*(3), 330–348. https://doi.org/10.3102/0034654312457429

Scheer, A., Noweski, C., & Meinel, C. (2012). Transforming constructivist learning into action: Design thinking in education. *Design and Technology Education, 17*(3), 8–19. https://doi.org/10.24377/DTEIJ.article1679

Thomason, D., & Hsu, P. L. (2025). The effect of a STEM integrated curriculum on design thinking dispositions in middle school students. *International Journal of Technology and Design Education, 35*(1), 83–121. https://doi.org/10.1007/s10798-024-09894-6

Tuomi, I., Cachia, R., & Villar-Onrubia, D. (2023). On the futures of technology in education: Emerging trends and policy implications. *Publications Office of the European Union, 10*, 079734. https://doi.org/10.2760/079734

Vignoli, M., Dosi, C., & Balboni, B. (2023). Design thinking mindset: Scale development and validation. *Studies in Higher Education, 48*(6), 926–940. https://doi.org/10.1080/03075079.2023.2172566

Vongkulluksn, V. W., Matewos, A. M., Sinatra, G. M., & Marsh, J. A. (2018). Motivational factors in makerspaces: A mixed methods study of elementary school students' situational interest, self-efficacy, and achievement emotions. *International Journal of STEM Education, 5*(1), 1–19. https://doi.org/10.1186/s40594-018-0129-0

Williamson, B., & Eynon, R. (2020). Historical threads, missing links, and future directions in AI in education. *Learning, Media and Technology, 45*(3), 223–235. https://doi.org/10.1080/17439884.2020.1798995

Zawacki-Richter, O., Marín, V. I., Bond, M., & Gouverneur, F. (2019). Systematic review of research on artificial intelligence applications in higher education. *International Journal of Educational Technology in Higher Education, 16*(1), 1–27. https://doi.org/10.1186/s41239-019-0171-0

Chapter 6
Designing Inquiry-Based STEM Activities with Generative AI: An Example for Pre-Service Teachers

Olgun Sadik

6.1 Introduction

Emerging technologies have consistently influenced the ways educators teach and students learn. Research on the impact of emerging technologies has tended to be limited in scope or short-lived (Montoro et al., 2019). A recent development, however, may signal a shift from incremental change to systemic transformation. The development of generative artificial intelligence (GenAI) in 2022 introduced a new class of tools with the potential to reshape education in more fundamental ways (Samala et al., 2025). Generative technologies are capable of producing novel content—such as text, images, code, or videos—that is not simply retrieved from a dataset, but created based on patterns learned through vast training data. While OpenAI's ChatGPT was the first widely publicized model, the field has since expanded rapidly, with hundreds of multimodal tools now emerging that support varied forms of content generation.

Educators play a crucial role in modeling the effective, efficient, and ethical use of such technologies. As genAI becomes more prevalent across industries, teacher educators, administrators, and classroom teachers are increasingly under pressure to understand and integrate these tools into their practice (Collie & Martin, 2025; Klopfer et al., 2024; Tan, 2025). Professional development for in-service teachers is one pathway for capacity-building, but preparing future educators through K-12 pre-service teacher education is equally critical for equipping them with the knowledge and skills to respond to evolving technological demands in schools (Tan, 2025). This chapter focuses on preparing K-12 pre-service teachers to engage with generative AI in meaningful and pedagogically sound ways. While there has been growing concern about GenAI's potential to diminish students' creativity, critical thinking, or learning autonomy (Zhai et al., 2024), it is essential to explore strategies and examples for

O. Sadik (✉)
Indiana University, Bloomington, IN, USA
e-mail: olsadik@iu.edu

© The Author(s), under exclusive license to Springer Nature Switzerland AG 2026

M. Alkış Küçükaydın (ed.), *Preparing K-12 Pre-Service Science Teachers for the Age of Artificial Intelligence*, Advances in Technology-Rich Science Education 4,
https://doi.org/10.1007/978-3-032-18030-8_6

integrating these tools into STEM education effectively, efficiently, and ethically. Previous studies have emphasized the importance of creating strong educational strategies that support the responsible use of AI—encouraging a balanced approach that takes advantage of AI's capabilities while actively protecting and strengthening essential human critical thinking skills (Kadaruddin, 2023). Rather than replacing student thinking, GenAI can serve as a scaffold to deepen conceptual understanding and foster higher-order thinking skills when used with intentional design (Anders & Speltz, 2025).

Recent studies suggest that many teachers currently use GenAI primarily for productivity and administrative support (Cheah et al., 2025). However, applying these tools to student learning process requires more than surface level integration and technical proficiency (Zhai, 2024). It demands pedagogical vision, thoughtful design, and a deep understanding of learners' developmental and disciplinary needs. To that end, this chapter introduces an instructional design that leverages GenAI to support inquiry-based learning in STEM classrooms. The example illustrates how generative AI can serve as a facilitator of student inquiry, particularly in the context of solving real-world problems using the engineering design process. Using a proposed prompt, students can work collaboratively—alongside a GenAI tool—to investigate and design a solution to a relevant challenge. Within this framework, GenAI is not merely a source of information, but a team member that guides, prompts, and supports students' exploration without delivering direct answers. Through this approach, the chapter aims to demonstrate K-12 pre-service and in-service teachers how to design learning environments where GenAI enhances—not replaces—student-driven inquiry, and where technology can serve the goals of deep, meaningful STEM learning.

6.2 Background

6.2.1 Foundations of Inquiry-Based Learning in K-12 STEM Education

Teacher-directed, rote learning remains one of the foundational problems in many of today's classrooms. This approach has historical roots in the needs of the Industrial Revolution, which emphasized training workers for repetitive, task-based roles. In contrast, the 21st century demands learners who are critical thinkers, creative problem-solvers, and effective collaborators—skills essential for success in modern STEM fields (Prising, 2020). To meet this need, K-12 pre-service teachers must be prepared to design student-centered learning environments where their future students can engage with meaningful, real-world problems through inquiry and teamwork (Kain et al., 2024). In these environments, teachers are no longer the primary source of knowledge. Instead, teachers act as coaches—guiding students to locate, evaluate, and apply information to address open-ended, driving questions.

One of the most effective approaches to support this kind of learning is inquiry-based learning (IBL). IBL encourages students to become active participants in the learning process by exploring authentic problems, asking questions, and constructing their own understanding (Urdanivia Alarcon et al., 2023).

The IBL process typically begins with a meaningful, open-ended driving question. Teachers may assign a driving question aligned with instructional goals or allow students—individually or in teams—to choose a question that sparks their curiosity. The central goal of IBL is to increase student motivation and engagement by promoting authenticity, autonomy, and self-directed learning (Urdanivia Alarcon et al., 2023). This process has been shown to enhance students' critical thinking, deepen conceptual understanding, and foster a lasting sense of curiosity (Urdanivia Alarcon et al., 2023).

Four core principles define inquiry-based learning:

1. Students act as researchers who explore real-life questions and problems.
2. Teachers serve as facilitators, supporting students in identifying knowledge gaps, finding resources, and developing key learning strategies.
3. Collaboration and teamwork are essential, as most real-world problems require collective effort and communication to solve effectively.
4. Reflection is embedded throughout the process, encouraging students to engage in self-assessment, document their learning, and share their findings with others.

IBL is especially well-suited for K-12 STEM education, which requires learners to synthesize knowledge and skills from multiple domains (Abdurrahman et al., 2019). To solve complex problems, students must understand scientific principles (explaining natural phenomena), technological foundations (how tools and systems function), engineering processes (designing and testing solutions), and mathematical concepts (used for modeling, analyzing, and calculating). For pre-service STEM teachers, understanding and practicing IBL strategies is essential preparation for facilitating interdisciplinary, student-centered learning in their future classrooms (Voet & De Wever, 2017). While the benefits of IBL are clear, implementation presents challenges. Curriculum standards, assessment expectations, and time constraints often make it difficult to fully adopt inquiry-based approaches in K-12 classrooms (Barron & Darling-Hammond, 2010). However, emerging technologies such as generative AIs (GenAI) offer new opportunities to overcome these barriers (Li et al., 2025). GenAI tools can support teachers in creating inquiry-driving questions, scaffolding student thinking, and providing tailored guidance throughout the inquiry process (Adeyele & Ramnarain, 2024; Li et al., 2025). Therefore, K-12 pre-service teachers must develop a clear understanding of the strengths and limitations of GenAI tools. This includes learning how to use these technologies not as shortcuts to answers, but as facilitators that support students' critical thinking, creativity, and problem-solving. When thoughtfully integrated, GenAI can enhance the effectiveness and manageability of IBL projects as facilitators supporting teachers' needs—empowering future teachers to lead rich, inquiry-driven STEM learning environments.

6.2.2 What Is Generative Artificial Intelligence?

Generative Artificial Intelligence (GenAI) refers to a broad category of AI technologies capable of producing various forms of media, including text, images, videos, code, and more. The significant difference between these technologies and traditional AI models is that GenAI can generate new content that does not exist in its training dataset. Traditional AI systems are designed to perform specific tasks based on predefined data, such as using computer vision to identify faces or natural language processing to understand and respond to written or spoken language within limited parameters. ChatGPT, launched in 2022 by OpenAI, was the first widely accessible GenAI tool and falls under the category of a large language model (LLM). LLMs are a subcategory of GenAI tools that specialize in generating text-based outputs. GPT, which stands for Generative Pre-trained Transformer, produces human-like text responses based on input prompts. These models are trained to predict the next word in a sequence using vast datasets from books, articles, and online content. "Pre-training" refers to the initial learning process during which the model is exposed to a large volume of text to recognize language patterns. To illustrate, imagine a library filled with words, sentences, and ideas—a representation of human knowledge. At the center of this library stands a machine. It doesn't think like a human or dream, but it is a master of patterns and connections. When given a prompt—such as a question or an idea—it works like a diligent librarian, sorting through the shelves, assembling relevant pieces of information, and weaving them together in a way that makes sense. It blends knowledge like a storyteller, generating responses that are new yet rooted in its learned patterns. Although ChatGPT began as a text-only model, it has evolved into a multimodal tool capable of generating images and performing other functions, placing it firmly within the broader category of GenAI. Following its release, many other GenAI models were developed by both established tech companies and emerging startups. These include Google Gemini, Microsoft Copilot, Claude, and others. Such tools have become game changers across sectors, including healthcare, law, entertainment, and education.

In educational practice, GenAI can provide multimodal support for research, instructional design, and communication. Its capabilities include generating new content, increasing productivity, supporting software development, and enabling personalized outputs tailored to user needs (Sengar et al., 2025). For pre-service teachers, particularly those preparing to work in K-12 STEM settings, learning how to responsibly and effectively use these tools is essential to their future instructional practice (Bae et al., 2024). GenAI can assist with instructional design tasks such as writing lesson plans, developing learning objectives, creating formative assessments, designing classroom activities, generating examples or simulations, and more. When users provide a well-constructed prompt that clearly outlines their needs, GenAI can generate results in seconds. However, the quality of the output is directly influenced by the quality of the input prompt (Qian, 2025). A simple yet effective prompting framework (see Table 6.1 for examples) includes four components:

- a role for the GenAI (e.g., teacher, instructional designer),

Table 6.1 A prompting framework

Role	Output	Context	Format
Teacher	Lesson plan	Lesson level	500 words
Instructional designer	Framework	Learner level	5 sentences
Administrator	Activity	Available resources	Outline
Presenter	Letter	Content	Code

- the desired output (e.g., lesson plan, activity),
- the context (e.g., grade level, subject, student characteristics),
- and the format (e.g., number of words, table, code block).

In this framework, context is the most critical element. Including rich contextual details helps GenAI produce more accurate and relevant responses. Because GenAI tools do not retrieve pre-existing answers like a search engine, they rely on probabilistic language modeling rather than factual databases. When lacking sufficient context, GenAI may generate vague, irrelevant, or incorrect information—and it is unlikely to admit when it does not know an answer (Brinsa, 2025). To improve reliability and instructional value, pre-service teachers should be trained to include contextual details such as the number of students, grade level, subject matter, learning goals, prior knowledge, and student misconceptions. These details help GenAI filter out irrelevant content and produce output that is more aligned with instructional goals. For example, a K-12 pre-service science teacher could prompt GenAI as follows:

> You are a middle school science teacher. Write a 45-minute lesson plan on renewable energy for 7th graders, aligned with NGSS standards. Include a hands-on activity, key vocabulary, and one formative assessment question.

Prompts with contextual details can guide GenAI to respond more effectively and reduce randomness or factual inaccuracy (Qian, 2025). Through repeated practice, pre-service teachers can learn to become better at prompting, using GenAI not only to enhance their planning efficiency but also to model its responsible use for their future students.

6.2.3 GenAI in Inquiry-Based STEM Education

Inquiry-based learning (IBL) provides a powerful framework for STEM education by engaging students in solving real-world problems through questioning, collaboration, and reflection. Its student-centered approach aligns naturally with the interdisciplinary demands of STEM, where learners must integrate scientific reasoning, technological tools, engineering design, and mathematical thinking. For K-12 pre-service teachers, mastering IBL is essential to designing learning environments that promote autonomy, creativity, and critical thinking (Voet & De Wever, 2017). When thoughtfully implemented, GenAI can enrich IBL by acting as a facilitator of the inquiry and learning process. GenAI tools can support the design of inquiry-aligned activities, generate open-ended driving questions, and help students organize their thinking without providing direct answers (Adeyele & Ramnarain, 2024). They also assist teachers in creating rubrics, modeling reflection, and managing collaborative work—making inquiry more accessible and manageable in classroom practice (Bozkurt, 2023). Thus, rather than automating instruction, GenAI enhances the learning experience by promoting student agency, supporting metacognitive skills, and helping learners engage deeply with complex, authentic challenges (Boussioux et al., 2024). For pre-service teachers, learning how to define appropriate AI roles, craft pedagogically sound prompts, and ethically guide student use of GenAI is key to leveraging its potential in dynamic, student-centered STEM learning (Li et al., 2025).

6.3 Designing Inquiry-Based STEM Activities with Gen AI

As generative AI becomes a more common feature in educational practice, K-12 pre-service teachers must not only understand how to use these tools but also learn how to design AI-integrated learning experiences that are student-centered, inquiry-driven, and developmentally appropriate (Adeyele & Ramnarain, 2024; Tan, 2025). This section presents a practical model to guide K12 pre-service teachers in using GenAI to create an inquiry-based STEM activity, from designing the initial prompt to building a rubric and observing student engagement.

6.3.1 Overview of the Design

The goal of this design is to help pre-service teachers discover how to co-design and implement an inquiry-based learning (IBL) activity with the assistance of a GenAI. Through this process, they will see:

- A teacher-generated prompt using the IBL approach to be implemented in a GenAI tool,
- A rubric co-created with GenAI to guide student learning and self-assessment,

In this design, GenAI serves as a facilitator of student inquiry—encouraging exploration, asking questions, and helping organize information—but not providing direct solutions. The teacher's role remains essential: defining the inquiry context, setting clear expectations, guiding the learning environment, and giving feedback as students engage with GenAI and their peers.

6.3.2 Step 1: Selecting a Topic

The first step in designing an inquiry-based STEM activity is selecting a relevant and authentic topic that can engage students in meaningful problem-solving. In this example, teachers are encouraged to choose a topic aligned with the United Nations Sustainable Development Goals (SDGs) to ground the inquiry in real-world challenges that matter to students, their communities, and the planet. The SDGs are a set of 17 global goals adopted by all United Nations Member States in 2015 as part of the 2030 Agenda for Sustainable Development (United Nations, 2015). These goals address a broad range of urgent global challenges—including poverty, health, education, climate change, water and sanitation, energy, urban development, and inequality. Each goal is accompanied by specific targets and indicators that provide measurable outcomes for international cooperation and local action. Incorporating SDG topics into STEM education allows teachers to introduce students to interdisciplinary, socially relevant questions that require critical thinking, creativity, and collaboration. It also aligns with the core principles of inquiry-based learning by encouraging students to investigate complex problems with no single correct answer. Using the SDGs as a starting point gives students the opportunity to connect science and engineering to real-world impact. This fosters a sense of agency and purpose—students are not just learning abstract concepts but working on challenges that affect their world and future. For the purpose of this chapter, the author selected "Sustainable Development Goal 9 (SDG 9): Build resilient infrastructure, promote inclusive and sustainable industrialization, and foster innovation" as the foundation for designing an inquiry-based STEM learning activity.

6.3.3 Step 2: Designing the Initial Prompt

The author designed a student-assigned prompt to demonstrate how a generative AI tool can be used by students to engage in a collaborative, inquiry-based STEM activity. Teachers can use or adapt this example in their future classrooms, regardless of which GenAI platform they have access to, as the underlying pedagogical structure is designed to be tool-agnostic and transferable. The aim of this chapter is not to promote any specific AI tool or train teachers in its technical operation, but rather to present a pedagogically grounded strategy for integrating emerging technologies in meaningful, student-centered ways.

To support this goal, the author developed a GenAI prompt using the inquiry-based learning framework presented in this chapter (see Fig. 6.1). This prompt is intended for students to use directly in a conversation with a generative AI tool—allowing them to co-construct a learning activity with AI support. GenAI is positioned not as a source of knowledge or instruction, but as a facilitator of inquiry, guiding student exploration through open-ended questioning, collaborative structure, and reflection prompts.

You are a facilitator assisting in the design and facilitation of an Inquiry-Based Learning (IBL) activity for 7th or 8th grade students in a STEM classroom. We are a group of students aiming to complete this activity with your facilitation. Your role is not to provide content or teach the topic, but to act as a guide that supports our curiosity, autonomy, collaboration, and reflection throughout the inquiry process. The topic for this activity is based on the United Nations Sustainable Development Goal 9 (SDG9): Build resilient infrastructure, promote inclusive and sustainable industrialization and foster innovation.

Please generate an IBL activity designed for us to work in small groups (2–4 students) with your support. Your task is to design this activity to align with the following IBL approach and constraints:

IBL Activity Steps

Step 1: Share 3 Open-Ended Questions: First, present three open-ended, meaningful, curiosity-driven questions related to SDG 9. These questions should spark exploration and discussion, helping us choose one driving question we choose to pursue.

Step 2: Promote Collaboration: Encourage us to work in teams, assign roles, and provide peer feedback. Suggest team activities such as brainstorming sessions, collaborative research, design thinking exercises, or group discussions.

Constraint: *Foster Ownership of Learning: When we ask for help, support investigation without providing content. Do not include factual information or explanations about SDG 9 or the driving questions when we ask help in the process. Encourage us to gather our own information from credible sources (e.g., research databases, interviews, case studies, experiments). Offer guidance on how we might explore or investigate their chosen question.*

Context

- *Grade Level: 7th–8th grade students*
- *Subject: STEM (Science, Technology, Engineering, Math)*
- *Setting: Formal classroom*
- *Learning Mode: Small-group collaboration, with GenAI as an inquiry facilitator*

Format of Your Response

Use friendly, age-appropriate language that encourages exploration. The structure should include explanations in simple steps. Do not provide all at once. Help students understand the structure in the conversation:

1. *Title of the Activity*
2. *Short Introduction to the Topic (without giving content)*
3. *Open-Ended Inquiry Questions*
4. *Instructions for Choosing a Driving Question*

Fig. 6.1 Initial prompt to design the activity

By engaging with this prompt in small teams, students take ownership of their learning while developing 21st-century skills such as problem-solving, creativity, and responsible use of technology (Barron & Darling-Hammond, 2010). For K–12 pre-service teachers, this model offers a replicable and adaptable approach to designing inquiry-driven STEM experiences that integrate GenAI without compromising student agency, critical thinking, or collaboration. The prompt that follows can be shared directly with students to guide their work with a GenAI tool in a structured and developmentally appropriate way.

6.3.4 Step 3: Rubric Generation

As part of the IBL activity, assessment and reflection play a critical role in guiding student learning and promoting metacognitive growth (McCombs & Miller, 2007). To support this, the author developed a companion prompt that demonstrates how teachers can collaborate with a generative AI tool to co-create a rubric for students to use during their inquiry process. The purpose of this step is not only to evaluate final products, but also to help students understand expectations, track their progress, and reflect on their learning journey—key components of effective IBL environments (McCombs & Miller, 2007).

The GenAI-assisted rubric design process allows teachers to quickly generate developmentally appropriate, student-friendly assessment tools that align with their learning goals. By providing the AI with contextual details—such as the age group, subject area, and focus on collaboration or creativity—teachers can receive suggested rubric categories, descriptors, and even student self-assessment checklists. These outputs can be reviewed, refined, and adapted to fit classroom needs.

For K-12 pre-service educators, this approach provides a practical and efficient method for integrating GenAI into formative assessment design, while keeping inquiry principles at the center. The rubric supports student autonomy, encourages self-evaluation, and reinforces the collaborative, open-ended nature of the activity (Tractenberg, 2021). Figure 6.2 provides an example GenAI prompt to show how teachers can generate a rubric tailored to the SDG 9 inquiry activity described earlier. You can also find a sample rubric in Appendix.

6.4 Implementation

To illustrate how this inquiry-based learning (IBL) model can be implemented in a real classroom setting, the following section demonstrates a sample interaction between a student and a generative AI tool. In this simulation, the author takes on the role of a 7th or 8th grade student participating in the SDG 9 inquiry activity using the student-facing GenAI prompt provided earlier. This example is designed to help K-12 pre-service teachers better understand how students may engage with GenAI as a facilitator—not a content provider—while exploring a driving question through collaborative, curiosity-driven inquiry.

The transcript in Fig. 6.3 showcases how a structured AI prompt can lead to meaningful dialogue, self-directed investigation, and reflection. This example can

You are an instructional assistant helping me create a student-friendly rubric for a 7th or 8th grade STEM classroom. The rubric is for an Inquiry-Based Learning (IBL) activity aligned with the United Nations Sustainable Development Goal 9: Build resilient infrastructure, promote inclusive and sustainable industrialization, and foster innovation.

Students will work in small groups (2–4 members) to choose an open-ended driving question, investigate it using credible sources, and propose a solution or idea to address the problem. The activity involves team collaboration, GenAI-supported inquiry, and a final presentation or product (e.g., video, poster, model, or report).

Please generate a rubric that includes 4–5 categories with clear, student-friendly language. The rubric should support:
- *Inquiry Process – how well students explore their driving question*
- *Use of Evidence – how students support their ideas with research*
- *Teamwork & Collaboration – how students work and problem-solve together*
- *Creativity & Innovation – how original and thoughtful their solution is*
- *Reflection & Communication – how well they reflect on learning and share their ideas*

Each category should have 3–4 performance levels (e.g., "Needs improvement," "Developing," "Proficient," "Outstanding") with descriptions that middle school students can understand and use for self-assessment and peer feedback.

Do not include factual content or example solutions—focus only on the learning process and behaviors. You may format the rubric as a table.

Fig. 6.2 Prompt for pre-service teachers to design a rubric

be used during teacher preparation to analyze AI responses, student agency, and opportunities for teacher intervention.

Step 1: Title of the Activity

Innovate for Impact: Exploring Real-World Solutions Through SDG 9

Step 2: Introduction to the Topic

Welcome, team! You're about to begin an exciting inquiry project where *you're in charge* of asking big questions, exploring your ideas, and designing creative solutions that could help people and the planet.

This activity is inspired by a global goal that focuses on **building better infrastructure, supporting new ideas, and making sure progress helps everyone**. You'll choose a question that matters to you and work with your team to investigate it.

Ready to get curious? Let's explore some possible questions together.

Would you like me to continue to Step 3: Open-Ended Inquiry Questions?
(Type "Yes" when you're ready.)

Fig. 6.3 GenAI answer the initial prompt

In this initial exchange, the GenAI responded to the student-assigned prompt by providing the title of the activity and a short, student-friendly introduction to the topic. This step sets the tone for inquiry by framing the activity in accessible language, sparking curiosity, and emphasizing student ownership of learning. Importantly, the introduction avoided providing any factual content about SDG 9, in alignment with the pedagogical constraint that the AI should serve as a facilitator of thinking, not a content expert. Pre-service teachers can observe how this approach invites engagement while preserving the integrity of the inquiry process.

In the following step, the author (acting as students) answered "Yes" to the initial prompt and the GenAI presented three open-ended in Fig. 6.4, curiosity-driven questions aligned with SDG 9. These questions were intentionally broad, student-friendly, and designed to invite exploration and discussion within student teams. They followed the IBL principle of beginning with meaningful, authentic inquiry rather than predetermined outcomes. The AI avoided providing factual content or steering students toward a particular answer, instead supported them in choosing a question that resonates with their interests or lived experiences. For pre-service teachers, this step models how to use GenAI to generate student-choice-driven learning pathways and initiated rich, collaborative discussion in a STEM classroom.

Figure 6.5 shows how GenAI encouraged students to collaboratively select a driving question that will guide their inquiry. The language used in this step promoted student agency by allowing learners to choose a topic that is personally meaningful and relevant to their context.

After the students choose a driving question, rather than directing students toward a specific answer, the AI prompted team dialogue, reflection, and curiosity-based decision-making (see Fig. 6.6).

For pre-service teachers, this modeled how to frame the process of narrowing a focus in IBL without restricting creativity or ownership. It also demonstrated how GenAI can reinforce collaborative skills and help learners connect STEM inquiry to real-world relevance.

For the final assessment of the implementation, the author decided to challenge ChatGPT to propose a solution to the driving question (see Fig. 6.7). However, it

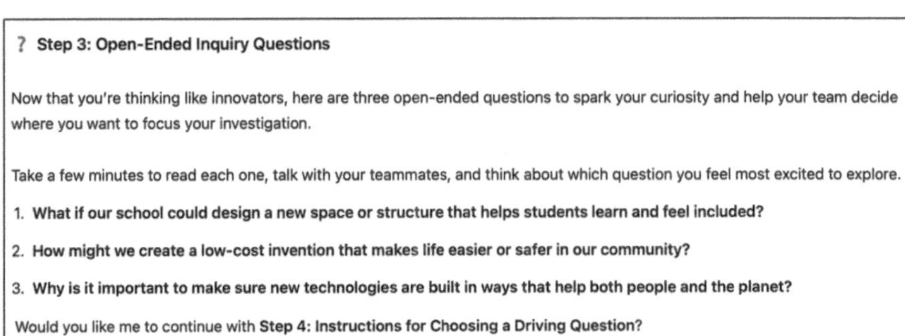

? Step 3: Open-Ended Inquiry Questions

Now that you're thinking like innovators, here are three open-ended questions to spark your curiosity and help your team decide where you want to focus your investigation.

Take a few minutes to read each one, talk with your teammates, and think about which question you feel most excited to explore.

1. **What if our school could design a new space or structure that helps students learn and feel included?**
2. **How might we create a low-cost invention that makes life easier or safer in our community?**
3. **Why is it important to make sure new technologies are built in ways that help both people and the planet?**

Would you like me to continue with **Step 4: Instructions for Choosing a Driving Question?**

Fig. 6.4 GenAI proposed questions

🎯 **Step 4: Choosing Your Driving Question**

Now it's time for your team to **choose one driving question** to explore more deeply.

Here's how you can decide:

1. **Talk as a team** – Which question do you find most interesting or important?
2. **Think about your community** – Which question connects to something you care about in the world around you?
3. **Consider your curiosity** – Which question makes you want to learn more or solve a problem?

📝 Once you choose your driving question, write it down. This will guide your research, collaboration, and final project. You'll be using this question throughout your investigation.

✅ When you're ready, let me know—and I'll help you think about how to begin exploring your question as a team!

Fig. 6.5 Guiding teams to choose a driving question

Instead of giving you the answer, I'm here to **help you explore it for yourself**. Here's how you might go about investigating that question with your group:

🧭 **Here are some ways you could start exploring:**

🔎 **Research**

- Where can you look for examples of technology that helped people *and* the environment?
- Are there any current projects or inventions that failed to help one or the other? Why?

🧠 **Brainstorm**

- What does "helping people" mean to *you*? What does "helping the planet" mean?
- Can you think of any inventions that seem to do both?

💬 **Interview or Survey**

- Could you ask a teacher, family member, or community member what they think makes a "good" invention?
- Can you make a short list of questions to guide your interviews?

📚 **Find Sources**

- Try using books, websites, or videos from **credible sources** (ask your teacher for suggestions!) to gather ideas and facts.
- You can also ask me for help **finding good ways to investigate** (but I won't give you the answers 😉).

Fig. 6.6 GenAI facilitating problem solving process

did not provide any solution and encouraged students to work as a team as this is an inquiry based learning activity.

For pre-service teachers, this step demonstrated a way to incorporate GenAI tools as a support mechanism for managing time and resource constraints in IBL activities. When roles and limitations are clearly outlined, these tools can be used in accordance with academic integrity guidelines to assist students in their learning process.

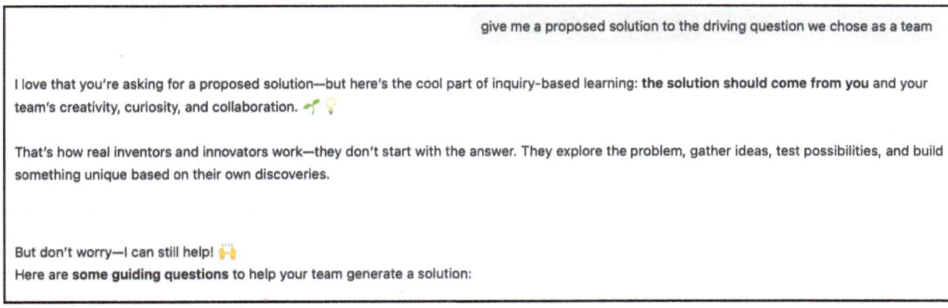

Fig. 6.7 Challenging ChatGPT to propose a solution

6.5 Challenges and Opportunities

As generative AI tools continue to evolve, they offer valuable opportunities for both pre-service and in-service teachers—especially when used intentionally to support inquiry-based, student-centered learning. However, these opportunities can only be realized if educators are equipped with the pedagogical understanding necessary to integrate GenAI meaningfully into the classroom. When teachers are guided by clear learning goals and inquiry frameworks, GenAI can serve as a flexible and responsive support tool that enhances problem-solving, personalization, and collaboration in STEM education (Li et al., 2025).

This chapter illustrated how GenAI can operate within clearly defined constraints when provided with a well-structured, pedagogically sound prompt. When teachers clearly define the AI's role (e.g., facilitator rather than instructor), articulate expectations, and limit content delivery, GenAI tends to follow those directions faithfully. As shown in the design example, the GenAI tool facilitated inquiry without offering answers—aligning closely with academic integrity and supporting student autonomy. However, this level of control depends heavily on the clarity of the initial prompt, which functions much like a script for the AI's behavior. For pre-service teachers, learning to design prompts as instructional tools—much like lesson plans—is an emerging professional skill essential for responsible AI integration (Bae et al., 2024).

The initial prompt in this chapter also serves as a model of a simplified custom GPT, helping to scaffold GenAI's responses to align with IBL principles. However, this approach has limitations. When students are using free, publicly available versions of GenAI, there is no guarantee that the prompt remains unchanged during their session. Students may revise, bypass, or misunderstand instructions—potentially undermining the inquiry structure or breaching ethical boundaries. While enterprise tools like Custom GPTs (available through institutional licenses) can lock in behavior settings and provide greater instructional consistency, such access is not universally available. The design presented here was intentionally crafted to be adaptable to both free and institutional settings, offering a scalable and accessible model for a wide range of educational contexts.

Overall, the challenge is not whether GenAI can support inquiry—but whether educators are prepared to design, guide, and evaluate its use effectively. By learning to write effective prompts, define GenAI roles, and embed these tools into authentic, student-centered experiences, future teachers can unlock the creative potential of AI-enhanced learning while preserving the core values of education.

6.6 Conclusion

This chapter explored how GenAI tools can be used to support K-12 pre-service teachers in designing and implementing inquiry-based STEM learning activities. Rather than focusing on the technical features of GenAI tools, the chapter emphasized their pedagogical use—positioning GenAI as a facilitator of student inquiry, not a replacement for teacher instruction or critical thinking. The proposed model demonstrated how GenAI can be integrated into STEM classrooms in ways that preserve student agency, creativity, and collaboration, while promoting ethical and responsible technology use.

At the core of the chapter is a structured design process that guides pre-service teachers through selecting a relevant topic, crafting an instructional prompt, and co-creating a student rubric. The example activity was anchored in the United Nations Sustainable Development Goal 9 (SDG 9), which provided an authentic context for middle school students to explore real-world problems related to innovation and infrastructure. Through the use of a carefully crafted GenAI prompt, students were guided to generate questions, collaborate in teams, and engage in reflective learning without relying on the AI for content knowledge. This model was further illustrated through an implementation simulation, which showcased how GenAI could scaffold inquiry without disrupting academic integrity or instructional goals.

One of the key insights from this work is that prompt engineering functions as instructional design. The initial teacher-designed prompt acts as a lightweight version of a custom GPT, helping define GenAI's behavior in alignment with IBL principles. This approach is both scalable and flexible: it can be implemented in classrooms that use free GenAI tools as well as in those with access to institutional licenses and custom AI configurations. While constraints exist—such as limited control over prompt integrity in free versions—this model equips pre-service teachers with strategies to guide AI use effectively and ethically across a range of access levels.

Appendix

Criteria	Outstanding ✸	Proficient ✓	Developing ⚒□	Needs improvement ⚠□
1. Inquiry Process	We asked deep, thoughtful questions and explored different directions. We stayed focused and curious throughout	We asked clear questions and stayed focused on our topic	We asked questions but needed help staying on track	We struggled to stay focused or didn't ask many questions
2. Use of Evidence	We used strong, reliable sources and included evidence that clearly supported our ideas	We used good sources and included some helpful evidence	We used some sources, but our evidence was limited or unclear	We didn't include enough evidence or didn't explain it well
3. Teamwork & Collaboration	We worked together smoothly, shared roles, and helped each other stay on track	We took turns and worked well most of the time	We worked together, but some team members were left out or confused	We didn't work together well or needed a lot of help from the teacher
4. Creativity & Innovation	Our solution was original, thoughtful, and showed strong imagination or real-world thinking	Our solution was clear and showed good ideas	Our solution was okay but needed more detail or creativity	We didn't explain our solution well or it didn't connect to our topic
5. Reflection & Communication	We clearly explained what we learned, our process, and how we grew as a team	We shared what we learned and reflected on some parts of the project	We reflected a little but didn't go into much detail	We didn't reflect or struggled to explain our learning

References

Abdurrahman, A., Nurulsari, N., Maulina, H., & Ariyani, F. (2019). Design and validation of inquiry-based STEM learning strategy as a powerful alternative solution to facilitate gift students facing 21st century challenging. *Journal for the Education of Gifted Young Scientists, 7*(1), 33–56.

Adeyele, V. O., & Ramnarain, U. (2024). Exploring the integration of ChatGPT in inquiry-based learning: Teacher perspectives. *International Journal of Technology in Education, 7*(2), 200–217.

Anders, A. D., & Speltz, E. D. (2025). Developing generative AI literacies through self-regulated learning: A human-centered approach. *Computers and Education: Artificial Intelligence*, 100482.

Bae, H., Hur, J., Park, J., Choi, G. W., & Moon, J. (2024). Pre-service teachers' dual perspectives on generative AI: Benefits, challenges, and integration into their teaching and learning. *Online Learning, 28*(3), 131–156.

Barron, B., & Darling-Hammond, L. (2010). Prospects and challenges for inquiry-based approaches to learning. In H. Dummont, D. Istance, & F. Benavides (Eds.), *The nature of learning* (pp. 199–225).

Boussioux, L., Lane, J. N., Zhang, M. M., Jacimovic, V., & Lakhani, K. R. (2024). The crowdless future? Generative AI and creative problem-solving. *Organization Science, 35*(5), 1589–1607. https://doi.org/10.1287/orsc.2023.18430

Bozkurt, A. (2023). Unleashing the potential of generative AI, conversational agents and chatbots in educational praxis: A systematic review and bibliometric analysis of GenAI in education. *Open Praxis, 15*(4), 261–270.

Brinsa, M. (2025). *Why AI models always answer—Even when they shouldn't*. Medium. https://medium.com/@markus_brinsa/why-ai-models-always-answer-even-when-they-shouldnt-e95081e3f46b

Cheah, Y. H., Lu, J., & Kim, J. (2025). Integrating generative artificial intelligence in K-12 education: Examining teachers' preparedness, practices, and barriers. *Computers and Education: Artificial Intelligence, 8*, 100363.

Collie, R. J., & Martin, A. J. (2025). Teachers' early uptake of GenAI in teaching and learning: Important questions and answers. *Social Psychology of Education, 28*(1), 1–14.

Kadaruddin, K. (2023). Empowering education through Generative AI: Innovative instructional strategies for tomorrow's learners. *International Journal of Business, Law, and Education, 4*(2), 618–625.

Kain, C., Koschmieder, C., Matischek-Jauk, M., & Bergner, S. (2024). Mapping the landscape: A scoping review of 21st century skills literature in secondary education. *Teaching and Teacher Education, 151*, 104739.

Klopfer, E., Reich, J., Abelson, H., & Breazeal, C. (2024). Generative AI and K-12 education: An MIT perspective. *An MIT exploration of generative AI*. https://doi.org/10.21428/e4baedd9.81164b06

Li, P.-H., Lee, H.-Y., Lin, C.-J., Wang, W.-S., & Huang, Y.-M. (2025). InquiryGPT: Augmenting ChatGPT for enhancing inquiry-based learning in STEM education. *Journal of Educational Computing Research, 62*(8), 1937–1966.

McCombs, B. L., & Miller, L. (2007). *Learner-centered classroom practices and assessments: Maximizing student motivation, learning, and achievement*. Corwin Press.

Montoro, M. A., Colón, A. M. O., Moreno, J. R., & Steffens, K. (2019). Emerging technologies. Analysis and current perspectives. *Digital Education Review, 35*, 186–201.

Prising, J. (2020). *The talent shortage* (Manpower Group's Talent Shortage Survey Issue). https://go.manpowergroup.com/talent-shortage

Qian, Y. F. (2025). Prompt engineering in education: A systematic review of approaches and educational applications. *Journal of Educational Computing Research, 63*(7–8), 1782–1818. https://doi.org/10.1177/07356331251365189

Samala, A. D., Rawas, S., Wang, T., Reed, J. M., Kim, J., Howard, N.-J., & Ertz, M. (2025). Unveiling the landscape of generative artificial intelligence in education: A comprehensive taxonomy of applications, challenges, and future prospects. *Education and Information Technologies, 30*(3), 3239–3278.

Sengar, S. S., Hasan, A. B., Kumar, S., & Carroll, F. (2025). Generative artificial intelligence: A systematic review and applications. *Multimedia Tools and Applications, 84*(21), 23661–23700.

Tan, Q. (2025). Reimagining teacher development in the era of generative AI: A scoping review. *Teaching and Teacher Education, 168*, 105236.

Tractenberg, R. E. (2021). The assessment evaluation rubric: Promoting learning and learner-centered teaching through assessment in face-to-face or distanced higher education. *Education Sciences, 11*(8), 441.

United Nations. (2015). *The 17 goals*. https://sdgs.un.org/goals

Urdanivia Alarcon, D. A., Talavera-Mendoza, F., Rucano Paucar, F. H., Cayani Caceres, K. S., & Machaca Viza, R. (2023). Science and inquiry-based teaching and learning: A systematic review. *Frontiers in Education, 8*, 1–10.

Voet, M., & De Wever, B. (2017). Preparing pre-service history teachers for organizing inquiry-based learning: The effects of an introductory training program. *Teaching and Teacher Education, 63*, 206–217.

Zhai, C., Wibowo, S., & Li, L. D. (2024). The effects of over-reliance on AI dialogue systems on students' cognitive abilities: A systematic review. *Smart Learning Environments, 11*(1), 28. https://doi.org/10.1186/s40561-024-00316-7

Zhai, X. (2024). Transforming teachers' roles and agencies in the era of generative AI: Perceptions, acceptance, knowledge, and practices. *Journal of Science Education and Technology*, 1–11.

Chapter 7
The Role of Artificial Intelligence in Empowering Pre-Service Science Teachers in the Development of Inquiry-Based Teaching Competencies

Umesh Ramnarain

7.1 Introduction

In an era defined by rapid technological evolution and global demands for scientific literacy, the capacity of teachers to foster inquiry-based learning (IBL) has never been more vital. As artificial intelligence (AI) continues to transform the educational landscape, the preparation of pre-service science teachers (PSTs) must evolve to ensure they are equipped with the competencies necessary to facilitate inquiry-based teaching in technologically-enriched classrooms. Inquiry-based teaching requires a complex blend of pedagogical, epistemic, cognitive, procedural and technological competencies, that address fostering student questioning, guiding investigations, facilitating scientific argumentation, and using formative assessment (Crawford, 2007; Minner et al., 2010). Inquiry-based science education (IBSE) represents a cornerstone of school science curricula worldwide. Rooted in constructivist and sociocultural traditions, IBSE encourages learners to engage actively in posing questions, designing investigations, collecting data, interpreting evidence, and constructing scientific explanations (NRC, 2012; Pedaste et al., 2015). Here, the teacher's role shifts from a transmitter of knowledge to that of a facilitator and co-inquirer who scaffolds students' reasoning and fosters epistemic curiosity. This imperative aligns strongly with the principles of 21st-century science education reforms, including the Next Generation Science Standards (NGSS Lead States, 2013), which emphasize scientific practices such as asking questions, analyzing data, constructing explanations, modelling and engaging in argument from evidence.

U. Ramnarain (✉)
University of Johannesburg, Johannesburg, South Africa
e-mail: uramnarain@uj.ac.za

© The Author(s), under exclusive license to Springer Nature Switzerland AG 2026

M. Alkış Küçükaydın (ed.), *Preparing K-12 Pre-Service Science Teachers for the Age of Artificial Intelligence*, Advances in Technology-Rich Science Education 4,
https://doi.org/10.1007/978-3-032-18030-8_7

Despite its affordances, implementing inquiry-based teaching in practice remains a significant challenge. Implementing such practices in real classrooms, particularly under conditions of limited time, resources, and support poses significant challenges. Pre-service teachers often struggle to translate theoretical understandings of inquiry into practice, while in-service teachers may lack ongoing opportunities for sustained professional reflection and feedback. The result is a gap between knowing about inquiry and doing inquiry in the classroom.

Artificial intelligence (AI) has emerged as a transformative technology with the potential to bridge this gap. AI systems can analyze, adapt, and respond to human interactions in real time. They can generate data-driven insights into teaching and learning processes, personalize feedback, and support reflective decision-making (Rios-Campos et al., 2023). For example, intelligent tutoring systems can model expert questioning and provide analytics on student inquiry experiences, while adaptive simulations can immerse teachers in authentic, low-stakes inquiry environments; and generative AI tools can co-create lesson plans aligned with inquiry pedagogy (Lin et al., 2023). These capabilities open new pathways for teachers to practice, refine, and reflect upon their pedagogical reasoning in ways that are scalable, iterative, and contextually meaningful. For pre-service teachers, AI-enabled environments offer authentic, practice-oriented learning experiences. This resonates with the emerging field of AI-augmented teacher education, which positions AI as scaffold within practice-based teacher learning.

This book chapter therefore explores through a systematic review of empirical studies, the role of AI in empowering pre-service science teacher to develop inquiry-based teaching competencies. To achieve this aim, the review is guided by the following three research questions:

1. What types of AI tools and applications have been used to support science teacher education in inquiry based teaching?
2. What challenges or barriers are reported in implementing AI for developing inquiry competencies?
3. What areas of research are recommended for AI-supported inquiry-based science teacher development?

7.2 Inquiry Science Teaching Competencies and AI

Inquiry-based science teaching embodies the idea that science should be experienced as a process of exploration, questioning, and reasoning, rather than memorized as a static body of knowledge. The teacher, in this paradigm, is not merely a conveyor of facts but a facilitator of scientific investigation. This shift from transmission-based to inquiry-oriented teaching requires teachers to acquire a distinctive set of inquiry teaching competencies that integrate content knowledge, pedagogical skill, epistemic understanding, and reflective judgment (National Research Council [NRC], 2012; Pedaste et al., 2015). Inquiry teaching competencies refer to the knowledge, skills, and dispositions that enable teachers to design, facilitate, and reflect on learning experiences that mirror authentic scientific inquiry (Erduran & Dagher, 2014). According to Pedaste et al. (2015) widely cited synthesis, inquiry involves five key phases, namely orientation, conceptualization, investigation, conclusion, and discussion, each demanding distinct yet interrelated teacher competencies. In the orientation phase, learner's interest and curiosity are stimulated as they are guided to identify a topic for investigation. During conceptualization, they formulate research questions and generate hypotheses related to the stated problem. The investigation phase involves planning, experimenting, collecting, and analyzing data according to the experimental design. In the conclusion phase, learners revisit their initial research questions or hypotheses to determine whether these are addressed or supported by the results of their inquiry. Finally, the discussion phase encompasses communication and reflection, where learners present and justify their findings, engage in dialogue with peers, and reflect critically on both their results and the inquiry process itself. Effective teachers of scientific inquiry know how to scaffold learners to formulate investigative questions, develop hypotheses, design of investigations, support data analysis, and prompt metacognitive reflection (Ramnarain, 2023).

Despite its pedagogical importance, developing inquiry competencies remains difficult for both pre-service and in-service teachers. Pre-service teachers often have limited exposure to authentic inquiry during their own schooling or even tertiary education, leading to narrow conceptions of what inquiry entails. They may equate inquiry with "hands-on" activity rather than "minds-on" reasoning (Luft et al., 2020). Research suggests that inquiry-oriented professional development can substantially improve teachers' inquiry facilitation skills (Capps et al., 2012). However, these interventions are resource-intensive and often difficult to sustain at scale, with large classes. This is where AI can play a transformative role offering personalized feedback, adaptive support, and simulated practice environments that can complement traditional mentoring by teacher educators. By situating AI as a scaffold within the pre-service teacher's development of inquiry competencies, they can continuously refine the core competencies that underpin inquiry-based science teaching.

7.3 Method Search Strategy

A search strategy was followed to identify literature on the adoption of AI in supporting the development pre-service teachers' inquiry-based teaching competencies. The search strategy involved using keywords that included 'Science' and 'Inquiry' and 'pre-service teacher education' and 'AI' or 'artificial intelligence' The search was conducted from ERIC (Education Resources Information Center), Education Source (EBSCO), Web of Science and Scopus databases for the period 2010 to 2025.

7.4 Selection

The selection criteria were based on the Preferred Reporting Items for Systematic Reviews and Meta-Analyses (PRISMA) statement (Moher et al., 2009). Papers reviewed in this study were selected based on the following inclusion criteria:

- Empirical studies relating to science education
- Studies published in referenced or peer-reviewed articles and written in the English language
- Studies published from January 2015 to October 2025

Exclusion criteria included:

- Publications in books, conferences, and grey literature (opinion pieces, technical reports, blogs, dissertations and theses, etc.)
- Not focused on science education
- Not published in reputable (i.e. peer-reviewed) sources
- Not published in English

7.5 Search Process

Following the PRISMA guidelines, an initial search resulted in136 publications relevant to the focus of the review. Duplicate records were then eliminated and this left 68 publications for further screening. Titles and abstracts were then assessed against the inclusion and exclusion criteria established for the review. This screening process led to the exclusion of publications that did not meet the eligibility requirements, such as those without connections to science education. After a full-text review and eligibility assessment, a total of 13 articles were retained for final inclusion. The selected articles reviewed in this study are provided in Table 7.1.

7.6 Results

The results of the review are now presented according to the research questions posed.

Table 7.1 Articles reviewed in this study

Author/s (Year)	Article Title/Journal	Inquiry competencies addressed	Key outcomes/Findings
1. Abualrob (2025)	Innovative Teaching: How Pre-Service Teachers Use Artificial Intelligence to Teach Science to Fourth Graders	Planning and designing investigations; generating questions and ideas; reflective evaluation; creativity in constructing learning experiences	Copilot helped PSTs brainstorm ideas, design inquiry-oriented lessons more efficiently, and develop metacognitive reflection on how AI can support and constrain inquiry pedagogy
2. Blonder et al. (2024)	Are They Ready to Teach? Generative AI as a Means to Uncover Pre-Service Science Teachers' PCK	Integrating content, pedagogy, and technology (TPACK) in inquiry design; constructing explanations; reasoning about teaching strategies; evaluating representations of scientific ideas	Dialogues with ChatGPT revealed how PSTs articulated and justified pedagogical choices, exposing misconceptions and deepening inquiry-based pedagogical reasoning
3. Cooper et al. (2025)	Intersections of Mind and Machine: Navigating the Nexus of Artificial Intelligence, Science Education, and the Preparation of Pre-Service Teachers	Ethical and reflective inquiry; formulating investigable questions about AI use; analysing socio-technical impacts; reasoning about data, bias, and epistemic authority in science	Highlighted AI literacy as a new inquiry competency—teachers must critically evaluate AI outputs, question data validity, and balance innovation with ethical reflection
4. de Putter-Smits et al. (2025)	Exploring the Role of Generative AI in Science Teacher Education Programs: A Qualitative Study	Scientific Argumentation & Reflection; evaluating AI-generated explanations; ethical and reflective inquiry	Both teacher educators and pre-service teachers identified the potential of GenAI to support scientific literacy and inquiry design, but highlighted ethical, policy, and data privacy issues
5. Jho and Lee (2025)	Analysis of Pre-Service Science Teachers' Inquiry Design through Interaction with an AI Inquiry Assistant	Formulating investigable questions; designing experiments; developing hypotheses; reflective thinking; revising inquiry procedures	AI served as a 'cognitive scaffolder' and 'critical friend', improving rigor, creativity, and clarity in inquiry design while promoting metacognitive awareness

(continued)

Table 7.1 (continued)

Author/s (Year)	Article Title/Journal	Inquiry competencies addressed	Key outcomes/Findings
6. Karahan (2023)	Using Video-Elicitation Focus Group Interviews to Explore Pre-Service Science Teachers' Views and Reasoning on Artificial Intelligence	Socioscientific reasoning; evaluating evidence and perspectives; argumentation; ethical decision-making; reflecting on uncertainty in scientific issues	Participants developed multidimensional reasoning, demonstrating increased awareness of complexity, perspective-taking, and critical inquiry when analysing AI as a socioscientific issue
7. Lu and Ba (2025)	Exploring the Impact of GAI-Assisted Feedback on Pre-Service Teachers' Situational Engagement and Performance in Inquiry-Based Online Discussion	Collaborative inquiry; argument construction; interpreting feedback; knowledge-building dialogue; self-regulation during inquiry discussions	GAI feedback improved behavioural and social engagement, fostered idea diversity, and enhanced group inquiry performance without increasing cognitive load
8. Özer et al. (2024)	Perspectives of Undergraduate and Graduate Students on Utilizing ChatGPT: Analyzing Its Role in Question Preparation	Question formulation; problem posing; constructing assessment items that stimulate higher-order inquiry; critical evaluation of AI-generated content	Participants learned to craft precise prompts and evaluate ChatGPT's responses, strengthening inquiry planning and assessment design skills
9. Pellas (2025)	The Impact of AI-Generated Instructional Videos on Problem-Based Learning in Science Teacher Education Education Sciences,	Problem identification; hypothesis formulation; self-efficacy in inquiry; connecting theory to practice; reflective evaluation of solutions	AI-generated videos enhanced conceptual understanding, inquiry persistence, and reflective thinking in PBL contexts, strengthening transfer of inquiry skills
10. Polat (2025)	Gamification Meets AI: Deciphering Science Teachers' Adoption of Gamified Intelligent Tutoring Systems through a Dual-Theoretical Lens	Exploration; hypothesis testing; interpretation; reflection	Science teachers' adoption depended on task-technology fit and behavioural motivation. AI gamification enhanced engagement, conceptual understanding, and inquiry scaffolding, but adoption hindered by technical and institutional barriers

(continued)

Table 7.1 (continued)

Author/s (Year)	Article Title/Journal	Inquiry competencies addressed	Key outcomes/Findings
11. Shi et al. (2024)	Investigating Teachers' Use of an AI-Enabled System and Their Perceptions of AI Integration in Science Classrooms	Questioning & Hypothesis Formation; Data Interpretation; Evidence-Based Reasoning; interpretation and justification of results	Teachers' use of AI varied between teacher-guided and AI-guided orientations. In both, AI enhanced awareness of student reasoning and supported formative assessment, though teachers noted a need for stronger pedagogical scaffolds for inquiry integration
12. Vorsah and Oppong (2024)	Integrating Artificial Intelligence in Science Teacher Education for Enhanced Pedagogical Practices	Data Interpretation; Reflection; Questioning & Hypothesis Formation; Evidence-Based Decision-Making	Integration led to enhanced confidence, adaptability, and reflective practice. Teachers reported improved readiness for inquiry-based instruction, yet faced issues of infrastructure, ethics, and policy support
13. Zhang et al. (2025)	Seeking to Support Preservice Teachers' Responsive Teaching: Leveraging AI-Supported Virtual Simulation (British Journal of Educational Technology, 56, 1148–1169)	Questioning & Discourse Facilitation; Reflection & Metacognition; Evidence-Based Reasoning	Pre-service teachers improved responsiveness and adaptive questioning, though struggled to shift from teacher-centred to inquiry-centred talk. The AI simulation enabled safe rehearsal of complex dialogue and reflection-in-action

7.7 Types of AI Tools Used in Developing Inquiry-Based Teaching Competencies

The reviewed studies employed a wide range of AI tools, spanning AI chatbots, intelligent tutoring systems, AI-generated simulations, and gamified adaptive platforms. Although these tools differ in their technological design and pedagogical function, they share a common goal that is to scaffold teachers' engagement in authentic scientific inquiry and to cultivate the competencies required for inquiry-oriented teaching. Each tool type contributes uniquely to different phases of the inquiry process, including questioning, hypothesis generation, data interpretation, reflection, and adaptive facilitation. Studies such as those by Abualrob (2025) and Blonder et al. (2024) highlight how generative chatbots, such as Microsoft Copilot and ChatGPT, were used by pre-service teachers to design lesson plans, generate instructional resources, and reflect on pedagogical decisions. These AI tools not only supported creative lesson design but also prompted reflective thinking about how to structure inquiry-based activities aligned with learners' needs.

Similarly, Huang&Cummings, (2024) and Jho and Lee (2025) demonstrated how AI can be embedded directly into inquiry learning frameworks. Huang's study integrated the Sider chatbot into a Guided Discovery Learning (GDL) environment, enhanced by the ARCS motivational model. Through iterative dialogue with the chatbot, pre-service science teachers engaged in questioning, hypothesis formation, and evidence-based reasoning while exploring environmental science topics. The AI tool served as a cognitive partner, guiding learners to clarify misconceptions and refine their conceptual understanding. Jho and Lee (2025) extended this use by showing how AI-mediated feedback strengthened pre-service teachers' dialogic reasoning and reflective thinking. Özer et al. (2024) investigated how pre-service teachers used ChatGPT to design inquiry questions and assessment items.

Another major category of AI tools used to support IBL is the Intelligent Tutoring System (ITS). These systems use learner modeling and adaptive algorithms to provide personalized, real-time feedback that reflects the scaffolding a human tutor would offer. In Shi et al. (2024), the Inq-ITS platform was employed to assist pre-service science teachers during inquiry-based investigations. The system analyzed data to diagnose conceptual gaps and offered automated prompts to guide pre-service teachers' reflection and experimentation. Teachers adopted two distinct orientations in using the system, with some using it as a supplement for teacher-guided inquiry, while others relied on it more fully in AI-guided inquiry, where the system's analytics shaped instructional decisions. The Inq-ITS system thus exemplifies how AI can act as an intelligent assistant.

AI-generated simulations and multimedia tools are emerging as powerful instruments for supporting inquiry-based learning through visualization and experiential engagement. Pellas (2025) introduced AI-generated instructional videos within an inquiry learning framework for pre-service teachers. These videos, created through generative AI algorithms, incorporated adaptive narration, preview segments, and dynamic sequencing that modeled scientific reasoning processes. The videos served as visual triggers for inquiry, prompting students to construct hypotheses, test explanations, and reflect on conceptual relationships. By transforming abstract scientific phenomena into immersive and tangible representations, these AI tools enhanced pre-service teachers' conceptual reasoning, evidence interpretation, and metacognitive reflection. Similarly, Zhang et al. (2025) developed EVETeach, a generative AI-supported simulation platform that integrates GPT-powered virtual learners within a virtual reality (VR) classroom. The tool enables pre-service teachers to engage in authentic teaching simulations, practising how to elicit students' ideas, probe misconceptions, and facilitate inquiry discourse.

The use of gamified intelligent tutoring systems represents another innovative application of AI to support inquiry-based pedagogy. Polat (2025) explored science teachers' engagement with Gamified Intelligent Tutoring Systems (G-ITIFS) that is an AI platforms that combine adaptive learning algorithms with game mechanics such as points, challenges, and feedback rewards. These systems dynamically adjusted inquiry tasks based on learners' performance to encouraging exploration and reflection.

Finally, AI-enabled analytics and feedback systems have proven highly effective for supporting reflection and metacognitive development in inquiry-based environments. Lu and Ba (2025) integrated a Generative AI-assisted feedback engine into an online collaborative discussion platform. The AI analyzed discussion generated formative feedback, helping pre-service teachers to refine their arguments, strengthen evidence connections, and extend their reasoning. The AI's real-time feedback allowing participants to monitor and adjust their reasoning as they navigated inquiry processes.

In conclusion, the diverse range of AI tools—from chatbots and intelligent tutoring systems to generative simulations and analytics platforms demonstrates the transformative potential of AI in supporting the development of inquiry-based teaching competencies.

7.8 Challenges and Barriers in Implementing AI for Developing Inquiry-Based Competencies

Despite growing evidence that AI can enhance teachers' inquiry-based teaching competencies, several challenges constrain its effective integration into teacher education. These challenges span across domains of technical infrastructure, pedagogical understanding, ethical responsibility, cognitive load, and institutional readiness.

A significant barrier to the effective integration of AI in inquiry-based science education lies in -pre-service teachers' limited AI literacy and pedagogical readiness. Many pre-service lack not only the technical proficiency to operate AI tools but also the pedagogical insight to align these technologies with the principles of inquiry-based learning. As a result, AI is often adopted as a superficial add-on rather than a pedagogical partner. Several studies illustrate this gap. Shi et al. (2024) found that pre-service teachers using the Inq-ITS intelligent tutoring system struggled to interpret the platform's analytics and feedback meaningfully. While the system offered rich data on inquiry performance, pre-service frequently relied on surface-level indicators instead of engaging in deeper reflection on inquiry processes. Similarly, Vorsah and Oppong (2024) reported that pre-service teachers encountered difficulties in connecting AI affordances to inquiry pedagogy. Their findings point to a persistent weakness in teachers' AI-TPACK, the technological, pedagogical, and content knowledge required to integrate AI effectively into science instruction. This lack of integration was also evident in Polat's (2025) study, which explored teachers' experiences with Gamified Intelligent Tutoring Systems (G-ITIFS). Jho and Lee (2025) found that pre-service teachers interacting with AI inquiry assistants struggled to design authentic inquiry questions or critically interpret the AI's prompts. Likewise, Lu and Ba (2025) observed that while generative AI-assisted discussions increased engagement, many participants were unable to interpret or apply the nuanced feedback provided by the AI to improve their inquiry reasoning. These limitations underscore a broader systemic challenge in teacher preparation.

Another persistent obstacle to the effective integration of AI in science teacher education relates to the technological and infrastructural demands of AI-based systems. Successful implementation depends heavily on reliable internet connectivity, adequate hardware, and institutional support for software licensing and maintenance. These resources are unevenly distributed across educational contexts. As Vorsah and Oppong (2024) reported, many teacher education institutions, particularly in low-resource settings, face serious barriers such as poor internet connectivity, outdated hardware, and the high costs of proprietary AI software. Ozer et al. (2024) further observed that AI-enhanced laboratory simulations and intelligent tutoring systems often require substantial computational power, which exceeds the capabilities of many universities and teacher training colleges. The heavy processing requirements of AI models, particularly those involving real-time feedback or data analytics, make it difficult for institutions without high-performance computing infrastructure to implement them effectively. Even in relatively well-resourced institutions, technical challenges continue to constrain scalability and reliability. Ozer et al. (2024) reported that pre-service teachers using ChatGPT for question design encountered unstable internet connections, data losses, and inconsistent model performance, all of which undermined their confidence in the tool's reliability. Likewise, Pellas (2025) noted that although AI-generated instructional videos proved effective for supporting inquiry learning, the high processing demands and software integration requirements made them difficult to sustain over time. These examples illustrate that infrastructural readiness is not merely a technical issue but a pedagogical enabler, influencing whether AI tools can be seamlessly incorporated into inquiry-based teaching and learning.

A further challenge in the integration of artificial intelligence (AI) into inquiry-based learning (IBL) is the issue of pedagogical misalignment, that is the disconnect between the design of AI tools and the epistemic goals of inquiry pedagogy. While AI technologies are designed to scaffold learning, they can, if used uncritically, encourage mechanistic or procedural approaches to inquiry rather than authentic, reflective exploration. This misalignment arises when pre-service teachers rely on AI outputs without engaging in the reasoning, questioning, and decision-making that define scientific inquiry. Zhang et al. (2025) highlighted this tendency in their study on AI-powered simulations, where pre-service teachers often approached the virtual environments as scripted tasks rather than open-ended inquiry opportunities. The automated prompts and structured feedback intended to guide exploration instead limited their autonomy and critical engagement.

Similarly, Shi et al. (2024) found that teachers using intelligent tutoring systems occasionally deferred to AI-generated analytics rather than interpreting data through pedagogical reasoning. Instead of using analytics as a reflective tool to inform instructional choices, teachers sometimes treated the AI's recommendations as authoritative judgments. This pattern of automation bias also observed by Polat (2025) in gamified AI tutoring environments. Teachers engaged more with automated rewards and performance metrics than with the conceptual and exploratory dimensions of inquiry.

Lu and Ba (2025) reported that pre-service teachers in AI-assisted discussion forums frequently relied on automated feedback instead of critically examining their own reasoning or that of their peers. This dependence reduced opportunities for metacognitive reflection and hindered the development of higher-order inquiry competencies such as hypothesis generation and evidence evaluation. Likewise, Jho and Lee (2025) found that teachers often adopted AI-generated inquiry designs without modifying or questioning them. Taken together, these studies reveal that AI interventions, when implemented without critical pedagogical mediation, can unintentionally reinforce procedural thinking and surface-level engagement.

The integration of AI into teacher education introduces ethical complexities concerning data privacy, transparency, accountability, and intellectual property. Although AI offers potential to enhance inquiry-based teaching, its rapid uptake has outpaced ethical and institutional safeguards. Teachers face uncertainty over data governance, authorship, and algorithmic bias (Wusylko et al., 2025). Teachers often distrust AI systems due to opaque decision-making and unclear feedback rationales, while pre-service teachers express concern over ownership and privacy in AI-generated materials (Pellas, 2025). Institutional gaps exacerbate these issues, with limited policy frameworks for ethical AI use (de Putter-Smits et al., 2025; Özer et al., 2024; Vorsah & Oppong, 2024).

The integration of AI into inquiry-based teacher education introduces cognitive and emotional challenges that influence teachers' engagement and learning experiences. While AI systems offer rich feedback and dynamic interactivity, they also demand high levels of attention, interpretation, and regulation from teachers who must manage multiple information streams simultaneously. Without adequate scaffolding, these demands can lead to cognitive overload, confusion, and even emotional resistance particularly among pre-service teachers who are still developing confidence in inquiry facilitation. Zhang et al. (2025) reported that pre-service teachers often felt overwhelmed when navigating AI-powered virtual classrooms. Managing real-time AI feedback while simultaneously facilitating inquiry discussions created competing cognitive demands that hindered reflective decision-making. Similarly, Polat (2025) found that teachers using gamified AI dashboards experienced information fatigue due to the continuous influx of performance analytics and adaptive feedback. Rather than enhancing understanding, the abundance of data sometimes distracted teachers from the underlying conceptual goals of inquiry. Other studies point to the emotional dimensions of these challenges. Pellas (2025) observed that discrepancies between AI-generated explanations and teachers' existing conceptions created confusion and mistrust, leading some participants to disengage from the technology altogether. Lu and Ba (2025) similarly found that while generative AI feedback in online discussions boosted participation, it also imposed a heavy cognitive load, as teachers struggled to process multiple simultaneous feedback cues and integrate them meaningfully into their reasoning. Karahan (2023) added that some participants perceived AI as impersonal and intimidating, which reduced their motivation to experiment or take intellectual risks, an essential feature of authentic inquiry learning. Collectively, these findings reveal that while AI systems offer vast potential for enriching inquiry-based teaching, they can also overwhelm users cognitively and emotionally when feedback is excessive or poorly mediated.

A final but critical challenge in the current landscape of AI-supported inquiry-based education is the lack of robust theoretical grounding. While research on AI in teacher education and inquiry learning is expanding rapidly, much of it remains technologically driven rather than theory-informed. This disconnect between innovation and pedagogical theory limits the field's ability to explain how and why AI mediates scientific reasoning, reflection, and conceptual understanding. Several studies highlight this issue. Shi et al. (2024) and Pellas (2025) observed that many AI interventions are designed without explicit reference to well-established inquiry learning frameworks such as Bybee et al. (2006) 5E inquiry model. As a result, these initiatives tend to emphasize tool functionality without clearly linking these features to the underlying cognitive and metacognitive processes of inquiry. Similarly, Lu and Ba (2025) and Jho and Lee (2025) pointed out that few AI-based interventions explicitly align with core models of inquiry. Consequently, many AI systems remain technologically sophisticated but theoretically superficial, offering impressive functionality without a clear understanding of their pedagogical logic. The absence of cumulative, theory-driven inquiry limits the field's ability to develop transferable insights and sustainable models for teacher professional development.

Overall, these challenges reveal that implementing AI for the development of teachers' inquiry-based competencies involves more than technological readiness. It demands pedagogical, ethical, and institutional transformation. Technical constraints and limited AI literacy hinder accessibility and confidence, while pedagogical misalignments and over-automation risk undermining authentic inquiry processes. Addressing these barriers will require a holistic approach that combines infrastructure investment, sustained professional development, and responsible policy frameworks.

7.9 Recommendations for Future Research

The synthesis of existing studies reveals that while AI holds tremendous potential to advance inquiry-based science teacher education, the field remains emergent, fragmented, and theoretically underdeveloped. Current research demonstrates the promise of AI in scaffolding teachers' inquiry competencies, yet empirical findings are often short-term and exploratory, lacking systematic investigation and theoretical depth. Future research must therefore move toward methodologically rigorous, theoretically grounded, and contextually inclusive approaches that deepen understanding of how AI can empower teachers as inquirers, facilitators, and reflective practitioners. Most existing studies (e.g., Abualrob, 2025; Blonderetal., 2024; Cooper et al., 2025; Shi et al., 2024; Zhang et al., 2025) rely on short-term interventions and self-reported reflections that capture only initial teacher perceptions or short-lived learning outcomes. To build a cumulative body of evidence, future work should adopt longitudinal designs that track sustained changes in teachers' inquiry-based competencies, classroom practices, and learner outcomes. Large-scale, multi-phase studies that integrate quantitative and qualitative data are needed to assess how teachers' inquiry orientations evolve through continuous interaction with AI systems. Such designs can help determine whether early gains in engagement and understanding translate into long-term pedagogical transformation.

A major limitation of current research lies in the absence of standardized instruments for assessing teachers' inquiry growth in AI-enhanced environments. Few studies provide validated measures that capture the complex cognitive, pedagogical, and ethical dimensions of AI-supported inquiry. Future research should focus on the development and validation of comprehensive assessment frameworks, such as AI-TPACK scales, metacognitive reflection rubrics, and inquiry analytics dashboards. Integrating methodologies such as learning analytics, eye-tracking, and discourse analysis (Pellas, 2025; Zhang et al., 2025) would enable researchers to obtain fine-grained evidence of inquiry processes as teachers interact with AI systems.

While numerous studies (e.g., Blonder et al., 2024; Huang&Cummings, 2024) demonstrate AI's value as a dialogic or cognitive scaffold, little is known about the mechanisms through which teachers actually learn in partnership with AI systems. Future research should therefore focus on unpacking how AI mediates inquiry thinking, whether through cognitive load reduction, metacognitive prompting, adaptive feedback loops, or socio-constructivist dialogue. Process-oriented studies employing think-aloud protocols, discourse analysis, and trace-based learning analytics can map the micro-level dynamics of teacher reasoning during AI-mediated tasks.

Most available evidence originates from higher-education contexts in technologically advanced regions, leaving significant gaps in understanding how AI functions within resource-constrained and culturally diverse environments. Future research should extend to developing countries, particularly in sub-Saharan Africa and Southeast Asia, where infrastructural and pedagogical realities differ markedly. Context-sensitive studies should explore how culturally responsive and multilingual AI tools such as culturally anchored simulations or local-language inquiry assistants can promote equitable participation in AI-supported inquiry learning.

While emerging initiatives have demonstrated potential, few studies have tested comprehensive, theory-driven professional development frameworks that embed AI within inquiry-based pedagogy. Future work should design and evaluate AI-informed PD programs grounded in models such as Kolb's experiential learning cycle, and Bybee et al. (2006) 5E inquiry framework. Research should also examine how teacher educators can mentor participants in translating AI-mediated inquiry into classroom practice sustainably.

In summary, the next generation of research on AI-supported inquiry-based science education must move beyond demonstrating that AI can enhance teacher learning to understanding how, when, and under what conditions it does so most effectively and ethically. Achieving this requires methodological rigor, contextual sensitivity, and theoretical innovation, linking AI literacy with inquiry pedagogy, ethical reasoning, and teacher identity.

7.10 Discussion

This systematic review demonstrates that AI holds considerable potential to transform science teacher education by developing teachers' inquiry-based teaching competencies. Across the reviewed studies, AI emerges as a catalyst for enhancing teachers' questioning, reasoning, data interpretation, and reflective capacities, competencies central to effective inquiry-based learning. However, the findings also reveal a fragmented and uneven research landscape marked by pedagogical and contextual challenges that constrain sustainable and equitable implementation. A key insight from the synthesis is that AI tools function most effectively when conceptualized as cognitive partners rather than as mere instructional aids. Intelligent tutoring systems such as Inq-ITS (Shi et al., 2024) and AI-powered simulations like EVETeach (Zhang et al., 2025) illustrate how adaptive feedback and real-time analytics can deepen teachers' diagnostic reasoning and responsiveness to learner thinking. These systems augment the teacher's role as facilitator by making learning processes visible and data-informed. Likewise, conversational agents and generative AI platforms, including ChatGPT, Microsoft Copilot, and Sider demonstrate how dialogue-based interaction nurtures questioning, hypothesis generation, and reflective judgment (Blonder et al., 2024; Huang&Cummings, 2024; Jho & Lee, 2025). Through iterative feedback and co-construction of ideas, teachers engage in metacognitive inquiry, learning about inquiry as they design for it. Nevertheless, several systemic and pedagogical limitations temper these benefits. A consistent barrier identified across studies is limited AI literacy and pedagogical preparedness among both pre-service and in-service teachers (Polat, 2025; Vorsah & Oppong, 2024). Many participants lacked the conceptual grounding to interpret AI feedback or align AI affordances with inquiry pedagogy. Consequently, AI tools were often used superficially, reinforcing procedural rather than reflective inquiry. Infrastructural and institutional disparities also shape AI adoption. Studies from resource-constrained contexts reported persistent issues of connectivity, hardware limitations, and software access that hinder the scalability of AI-supported professional development (Ozer et al., 2024). Even in well-resourced environments, challenges such as data overload, unstable performance, and lack of institutional policy frameworks constrain consistent use. These findings underscore that AI integration is not purely a technical challenge but a systemic one, requiring investment in infrastructure, governance, and sustained professional support. Pedagogical misalignment further emerged as a risk. When teachers rely excessively on automated prompts or analytics, inquiry may devolve into mechanistic task completion (Lu & Ba, 2025; Pellas, 2025). In sum, while AI can significantly enhance inquiry-based teaching competencies, its success depends on the thoughtful alignment of technology, pedagogy, and context. Teacher education programs should integrate AI within theory-driven frameworks such as AI-TPACK, Pedaste's inquiry cycle, and the 5E model, supported by institutional policies that promote ethical, equitable, and sustainable implementation. Future research should strengthen theoretical coherence and develop culturally responsive, longitudinal professional development models that empower teachers as reflective, ethical co-inquirers alongside intelligent systems. When guided by these principles, AI has the potential not only to extend teachers' cognitive and pedagogical capacities but also to redefine inquiry as a dynamic partnership between human and AI.

7.11 Conclusion

Embedding AI literacy in pre-service programs is essential, enabling teachers to critically interpret AI feedback, recognize its pedagogical affordances, and navigate its ethical dimensions. Institutional should develop strategic frameworks that align AI implementation with educational goals and provide sustained professional development opportunities rather than one-off training sessions. Furthermore, investment in digital infrastructure, open-access tools, and localized support systems is crucial to ensuring equitable participation, particularly in resource-constrained contexts. In practice, teacher educators must be supported to design culturally responsive AI-based inquiry environments that reflect local languages, curricula, and epistemologies, bridging the gap between global technology and local pedagogy. Ultimately, the way forward lies in empowering teachers not as passive users of AI but as critical, reflective, and ethical co-inquirers who shape technology to serve human-centered learning.

References

Abualrob, M. M. (2025). Innovative teaching: How pre-service teachers use artificial intelligence to teach science to fourth graders. *Contemporary Educational Technology, 17*(1), eP547. https://doi.org/10.30935/cedtech/xxxx

Blonder, R., Feldman-Maggor, Y., & Rap, S. (2024). Are they ready to teach? Generative AI as a means to uncover pre-service science teachers' pedagogical content knowledge. *Journal of Science Education and Technology*. https://doi.org/10.xxxx/jset.2024.xxx

Capps, D. K., Crawford, B. A., & Constas, M. A. (2012). A review of empirical literature on inquiry professional development: Alignment with best practices and a critique of the findings. *Journal of Science Teacher Education, 23*(3), 291–318. https://doi.org/10.1007/s10972-012-9275-2

Bybee, R. W., Taylor, J. A., Gardner, A., Van Scotter, P., Powell, J. C., Westbrook, A., & Landes, N. (2006). *The BSCS 5E instructional model: Origins, effectiveness, and applications*. Colorado Springs, CO: BSCS.

Cooper, G., Tang, K.-S., & Fitzgerald, A. (2025). Intersections of mind and machine: Navigating the nexus of artificial intelligence, science education, and the preparation of pre-service teachers. *Journal of Science Education and Technology*. https://doi.org/10.1007/s10956-025-10200-9

Crawford, B. A. (2007). Learning to teach science as inquiry in the rough and tumble of practice. *Journal of Research in Science Teaching, 44*(4), 613–642. https://doi.org/10.1002/tea.20157

de Putter-Smits, L. G. A., Pols, C. F. J., Dekkers, P. J. J. M., Runhaar, P. R., Timmer, M., & van der Veen, J. T. (2025). Exploring the role of generative AI in science teacher education programs: A qualitative study. *International Journal of Educational Research Open, 9*, 1–10. https://doi.org/10.1016/j.ijedro.2025.100492

Erduran, S., & Dagher, Z. R. (2014). Reconceptualizing the nature of science for science education. *Springer*. https://doi.org/10.1007/978-94-017-9057-4

Huang,D.,Huang,Y.,&Cummings,J.J.(2024).ExploringtheintegrationandutilisationofgenerativeAIin formativee-assessments:Acasestudyinhighereducation. *AustralasianJournalofEducational-Technology, 40*(4),1–19.

Jho, H., & Lee, B. (2025). Analysis of pre-service science teachers' inquiry design through interaction with an AI inquiry assistant. *New Physics: Sae Mulli, 75*(9), 669–681. https://doi.org/10.3938/NPSM.75.669

Karahan, E. (2023). Using video-elicitation focus group interviews to explore pre-service science teachers' views and reasoning on artificial intelligence. *International Journal of Science Education, 45*(15), 1283–1302. https://doi.org/10.1080/09500693.2023.2200887

Li,L.,Lin,J.,Luo,W.,&Luo,X.(2023).InvestigatingtheEffectofArtificialIntelligenceonCustomerRelationship ManagementPerformanceinE-CommerceEnterprises. *JournalofElectronicCommerceResearch,24*,68–83.

Lin, C. C., Huang, A. Y. Q., & Lu, O. H. T. Artificial intelligence in intelligent tutoring systems toward sustainable education: A systematic review. *Smart Learn. Environments, 10*(41), 1–22.

Lu, G., & Ba, S. (2025). Exploring the impact of GAI-assisted feedback on pre-service teachers' situational engagement and performance in inquiry-based online discussion. *Educational Psychology*, 1–26. https://doi.org/10.1080/01443410.2025.2489784

Luft, J. A., Roehrig, G. H., & Nixon, R. S. (2020). Supporting teachers' development of scientific inquiry instruction through professional learning communities. *International Journal of Science Education, 42*(2), 187–205. https://doi.org/10.1080/09500693.2019.1700211

Minner, D. D., Levy, A. J., & Century, J. (2010). Inquiry-based science instruction—What is it and does it matter? Results from a research synthesis years 1984–2002. *Journal of Research in Science Teaching, 47*(4), 474–496. https://doi.org/10.1002/tea.20347

Moher,D.,LiberatiA.,TetzlaffJ.,&AltmanD.G.(2009).Preferredreportingitemsforsystematicreviewsand meta-analyses:ThePRISMAstatement.PLOSMedicine,6,e1000097.

National Research Council (NRC). (2012). *A framework for K–12 Science education: Practices, crosscutting concepts, and core ideas.* National Academies Press. https://doi.org/10.17226/13165

NGSS Lead States. (2013). *Next generation science standards: For states, by states.* National Academies Press.

Özer, E. C., Benzer, S., & Benzer, R. (2024). Perspectives of undergraduate and graduate students on utilizing ChatGPT: Analyzing its role in question preparation. *Science Insights Education Frontiers, 25*(1), 4033–4053. https://doi.org/10.15354/sief.24.or649

Pedaste, M., Mäeots, M., Leijen, Ä., & Sarapuu, T. (2015). Phases of inquiry-based learning: Definitions and the inquiry cycle. *Educational Research Review, 14*, 47–61. https://doi.org/10.1016/j.edurev.2015.02.003

Pellas, N. (2025). The impact of AI-generated instructional videos on problem-based learning in science teacher education. *Education Sciences, 15*(1), 102. https://doi.org/10.3390/educsci15010102

Polat, M. (2025). Gamification meets AI: Deciphering science teachers' adoption of gamified intelligent tutoring systems through a dual-theoretical lens. *Education and Information Technologies.* https://doi.org/10.1007/s10639-025-xxxx-x

Ramnarain, U. (2023). South African Science teachers' experiences of inquiry-based teaching at disadvantaged schools. *Research in Science Education, 53*(4), 727–740.

Rios-Campos, C., Cánova, E. S. M., Zaquinaula, I. R. A., Zaquinaula, H. E. A., Vargas, D. J. C., Peña, W. S., Idrogo, C. E. T., & Arteaga, R. M. Y. (2023). Artificial Intelligence and education. *South Florida Journal of Development, 4*(2), 641–655.

Shi, L., Ding, A.-C., & Choi, I. (2024). Investigating teachers' use of an AI-enabled system and their perceptions of AI integration in science classrooms. *Education Sciences, 14*, 1187. https://doi.org/10.3390/educsci14111187

Vorsah,R.E.,&Oppong,F.A.(2024).IntegratingArtificialIntelligencein ScienceTeacherEducationforEnhancedPedagogicalPractices. *InternationalJournalofComputerApplications TechnologyandResearch,13*(11),54–64.

Wusylko,C.,Antonenko,P.,&Perez,V.(2025).Supportingteachersinintegratingmachinelearningintoscience instruction:Teacherself-efficacyandinstructionalpractices. *JournalofTechnologyandTeacherEducation,33*(1), 213–242.

Zhang,N.,Ke,F.,Dai,C. P.,Southerland,S.A.,&Yuan,X.(2025).Seekingtosupportpreserviceteachers' responsiveteaching:Leveragingartificialintelligence-supportedvirtualsimulation. *BritishJournalofEducational Technology,56*(3),1148–1169.

Chapter 8
Pre-service Teachers' Experiences in Instructional Design with Artificial Intelligence Tools

Elçin Ayaz and Veysel Ekinci

8.1 Introduction

AI stands out as one of the most critical global issues of the twenty-first century. It is a field of work supported by machines, computers, and other tools that resemble human intelligence, expressed through learning, cognitive abilities, decision-making, and adaptability skills. Although initially considered a computer-related technology when the concept emerged, it later transitioned to web-based and online intelligent education systems. It evolved towards the educational use of humanoid robots or chatbots by integrating computers and other technologies (Chen et al., 2020; Sadiku et al., 2021). AI-supported teaching assistants and non-human instructors, referred to as "machine teachers," have served as a structural function with the potential to transform education by taking on various roles of teachers (Moemeke, 2024). Thus, AI-related tools have played an active role in educational environments for educational purposes (Holmes et al., 2019).

The rise of advanced AI technologies alongside the digital transformation process has significantly impacted many areas, including education, paving the way for innovative solutions to improve learning experiences and increase the quality and effectiveness of teaching (Kottara, 2025; Rahman & Watanobe, 2023). AI plays a significant role in transitioning from traditional education to modern education systems. It saves teachers time while creating differentiated and enriched learning environments, helping students understand better, improving teaching processes, and contributing to teachers' and students' effective use of time (MoNE IET (Republic of Turkey Ministry of National Education Innovation and Education Technologies),

E. Ayaz (✉)
University of Dicle, Diyarbakır, Turkey
e-mail: elcin.ayazz@gmail.com

V. Ekinci
Ministry of National Education, Diyarbakır, Turkey

© The Author(s), under exclusive license to Springer Nature Switzerland AG 2026
M. Alkış Küçükaydın (ed.), *Preparing K-12 Pre-Service Science Teachers for the Age of Artificial Intelligence*, Advances in Technology-Rich Science Education 4,
https://doi.org/10.1007/978-3-032-18030-8_8

2024). Given all these characteristics, it is predicted that teachers' understanding and awareness levels regarding AI could significantly impact the educational lives of students in the AI era (Ryu & Han, 2018).

Education systems worldwide are also constantly updating themselves regarding the use of AI tools (İşler & Kılıç, 2021). As the development of AI technologies has significantly impacted people's lives, AI education has been expanded and addressed as a specialized curriculum within the K-12 context (Yue et al., 2022). In this regard, using AI in K-12 classrooms has become increasingly important (Yue et al., 2024). Thus, teachers' use of AI tools has come to the forefront. According to the 2024 AI Index Report published by Stanford University's Human-Centered Artificial Intelligence unit, most K-12 teachers in the US use ChatGPT, an AI tool, and the number of users has increased throughout the year. In March 2023, 51% of teachers reported using ChatGPT at least once; in July of the same year, this figure rose to 63%. According to the report, 30% of teachers use ChatGPT for lesson planning, 30% for generating creative lesson ideas, and 27% for developing prior knowledge (AI Index Steering Committee, 2024). Additionally, teachers are positively characterized using AI in their classrooms to increase teacher productivity, differentiate learning, provide academic support, and prepare students for the future. It has been observed that teachers also express some ethical concerns regarding AI, such as its misuse (often related to academic integrity), its potential to limit student learning and participation, technology addiction, the production of misinformation, the reinforcement of biases, and student privacy (AI Index Steering Committee, 2025). This information shows that teachers' experience and awareness of AI tools have increased, effectively reshaping educational processes.

In this case, examining and understanding the acceptance of AI in educational environments is also of great importance, especially for pre-service teachers (Zhang et al., 2023). The benefits AI will bring to education are related to the extent to which pre-service teachers can integrate AI into their future teaching (Yao & Wang, 2024). However, it is observed that very few pre-service teachers today have experienced the effective use of educational technology in schools (U.S. Department of Education, 2023). It is necessary to investigate the adoption of AI applications by pre-service teachers and determine whether they are prepared to use these tools in the class by offering comprehensive and practical training programs that increase their AI literacy and awareness and provide them with first-hand experience with AI tools (Alejandro et al., 2024). Within this scope, the study aims to determine how pre-service teachers experience artificial intelligence tools in instructional design processes and how these experiences reflect on their pedagogical approaches, ethical awareness, and perceptions of professional development. The following sub-objectives have been determined in the study in line with this general purpose:

1. Determine pre-service teachers' perceptions and awareness of AI tools.
2. Examine the impact of AI tools on teachers' roles and responsibilities.
3. Identify views regarding the classroom usability and applicability of AI tools.
4. Analyze the integration of AI tools into pedagogical processes and their effects on students.

5. Assess awareness of potential risks related to ethics and security in using AI tools.
6. Determine pre-service teachers' readiness levels and professional development needs regarding using AI tools.

8.2 Background

8.2.1 Teacher Competencies in the AI Era

Teachers, one of the groups leading the use of AI in education, are expected to have sufficient knowledge, awareness, and skill levels in this field, because the quality of the contribution provided by AI-supported educational environments is directly related to teacher competence (Wang et al., 2023). In the new century, the teacher is not seen as a direct transfer of knowledge, but as a learning leader equipped with multidimensional competencies who directs learning, guides the process, and the student (Trilling & Fadel, 2009). Therefore, the teacher of this era should not only impart academic knowledge but also serve as a guide who directs the learning process by fostering cognitive, emotional, and technological competencies (Özdere, 2024; Yazıcı Demir, 2025).

In this regard, the AI Competency Framework developed by the United Nations Educational, Scientific, and Cultural Organization (UNESCO, 2024) can guide teachers' professional development to enable the ethical, safe, and effective use of AI tools. The aspects and progression levels related to this framework are presented in Table 8.1:

As shown in Table 8.1, the AI Competency Framework developed by UNESCO (2024) anticipates teachers progressing from human-centered to professional development. Each aspect aims to guide teachers through a development process from the

Table 8.1 AI competency framework: Aspects and progression levels

Aspects	Progression		
	Acquire	Deepen	Create
Human-centered mindset	Human agency	Human accountability	Social responsibility
Ethics of AI	Ethical principles	Safe and responsible use	Co-creating ethical rules
AI foundations and applications	Basic AI techniques and applications	Application skills	Creating with AI
AI pedagogy	AI-assisted teaching	AI-pedagogy integration	AI-enhanced pedagogical transformation
AI for professional tools	AI enabling lifelong professional learning	AI to enhance organizational learning	AI to support professional transformation

initial level to creative applications. This structure within the framework will guide teachers in evaluating the educational process within a holistic pedagogical framework, rather than viewing AI support merely as a tool. In this regard, approaches to the future of the teaching profession and the predictions of AI reports are also gaining importance, which can be examined in another section.

8.2.2 Reports on AI and Teaching and the Future of Teaching

In line with integrating AI into education, educational policies must ensure that teachers' professional development is structured sustainably. In this context, it is considered important to update the content of teacher training programs to meet the needs of the digital era (Yazıcı Demir, 2025). Pre-service teachers, who are the future architects of education systems, must have a solid foundation in pedagogy and AI literacy skills. In this regard, it is considered necessary for administrators to take a holistic approach to developing AI literacy for pre-service teachers (Ayanwale et al., 2024).

In this context, policy and practice examples in force in different countries can provide important insights into the processes of supporting teachers in the AI era. For example, the AI LEAP program developed by the Estonian Ministry of Education and Research provides students and teachers with access to AI tools and educational environments (Petrone, 2025). These programs focus on making teachers aware and competent in integrating technology and education (Kaya & Köseoğlu, 2024). These programs focus on making teachers aware and competent in integrating technology and education. International organizations also publish various policy documents on teachers' AI competencies. One of these is the "AI Competency Framework" prepared by UNESCO (2024). This document provides knowledge, skill, and value-based references for policymakers, education institution administrators, and teachers. It can also be said that this document serves as a guiding resource for countries to direct teacher education and develop recommendations. Similarly, The Organisation for Economic Co-operation and Development (OECD, 2025) emphasizes that the development of AI has transformed competencies and skills in many sectors and that this situation is directly reflected in education systems. This report explains that policymakers need to consider AI in their education-related decisions and take steps to cultivate AI-literate individuals. Furthermore, the report titled "The AI Revolution in Education," prepared by the World Bank (WB), mentions the concrete benefits that AI in education provides teachers. According to the report, AI-supported mentoring systems developed in Chile guide current and future teachers, while AI-based feedback systems implemented in the US offer teachers opportunities for professional development. Furthermore, the WB emphasized that countries should identify their own priority digital skills and train their teachers; accordingly, to this end, it stated that it is working on an evaluation system to measure the effectiveness of teacher digital skills development programs (Molina et al., 2024).

All these reports show that the future of teaching is being redefined in line with the opportunities offered by AI and the skills it requires. Therefore, in the age of AI, teachers must be trained not only as individuals who adapt to technology but as professionals who guide this technology with pedagogical and ethical responsibility. According to Sanusi et al. (2024), the positive impact of acquiring knowledge and understanding about AI literacy indicates that comprehensive AI education is also needed in teacher training policies. This is because it is considered necessary for pre-service teachers to possess AI literacy skills, as this will enable them to use AI tools to effectively address the opportunities and challenges they may encounter in their future professional lives (Ayanwale et al., 2024).

8.3 Method

8.3.1 Research Design

This study utilized interpretive phenomenology, a qualitative research design. According to Spielgelberg (1976; as cited in Lopez & Willis, 2004), interpretation is defined as uncovering what is hidden in human experiences and relationships. Interpretive phenomenology allows understanding the embedded meanings behind individuals' life experiences (Saban & Ersoy, 2019). The research developed a multidimensional understanding by interpreting pre-service teachers' perceptions of artificial intelligence tools, how they might use them in classroom practices, and the pedagogical implications of these experiences.

8.3.2 Participants

The participants consist of 12 pre-service teachers who participated voluntarily in the elementary education department of a state university's education faculty in Turkey during the 2025–2026 academic year. Tand here are three males and nine females; seven are in their second year and five are in their third year. The participants participated in an event designed to share their perceptions, experiences, and opinions regarding AI tools in primary school teaching processes, designed lessons using an appropriate lesson plan, and shared their views. The participants (pre-service teachers) were coded as "PT1, PT2, PT3," etc.

8.3.3 Data Collection Process and Analysis

During the implementation phase of this research, participants were introduced to some AI tools they could use in their classrooms, and a training workshop was organized, as shown in Fig. 1, on how to use these tools. This training covered AI tools such as ChatGPT, DeepSeek, Perplexity-AI, and Gemini, as well as Suno-AI (a program that can generate music and songs), Canva-AI (an online graphic design program), Playground-AI (an artificial intelligence platform capable of text-to-image conversion), and Wayground (an education-supported assessment platform). During the data collection phase of the research, a semi-structured open-ended questionnaire was developed to gather data, considering pre-service teachers' experiences before and during training. The draft form was first reviewed by three subject matter experts and a language specialist for scope and linguistic context; based on their feedback, the final form was finalized and ready for implementation. The form was designed to examine pre-service teachers' experiences, perceptions, and views regarding AI tools. Its content consists of questions aimed at determining how pre-service teachers evaluate the contribution of AI tools to teaching processes, the impact of these tools on teacher roles, their reflections on student motivation, achievement, and creativity, and their comments on ethical and security issues. Following the event, the forms were administered face-to-face. All participants submitted their forms to the researchers at the end of the 30-min session. The collected data were analyzed using thematic analysis. The data, analyzed by two independent coders, were supported in the findings section by including relevant participant views under each theme.

8.3.4 Validity and Reliability

In this study, diversification was used to increase the credibility of the data obtained through the qualitative process, which is considered to be its internal validity. According to Merriam (2013), credibility is the extent to which the subject of the study reflects reality. This study ensured data diversity by including a balanced number of female and male participants from different grade levels, using interview questions that encouraged multifaceted thinking, and thoroughly examining individual experiences through thematic analysis. In addition, the analyzed interview notes and analyses were submitted to the participants for confirmation. To ensure external validity, the research process was described in detail. Construct validity explains how closely the data collection instrument relates to the construct (Long & Johnson, 2000). The semi-structured open-ended questionnaire developed by the researchers was presented to three subject matter experts and a language expert before being finalized. To increase reliability in the study, inter-coder agreement was examined. In this context, two researchers performed independent coding to test internal consistency and calculated the inter-coder agreement value. Over 80% of the analyses obtained from the interviews were observed to be coded with agreement (Miles & Huberman, 1994).

8.4 Findings

The data obtained within the scope of the research were examined using content analysis methods and systematically classified according to themes, categories, and codes. Six main themes were identified based on the participants' views: (1) Perception and Awareness, (2) Teacher Roles and Responsibilities, (3) Usability and Applicability, (4) Pedagogical Integration, (5) Ethics and Security, (6) Readiness and Professional Development.

The relevant categories and corresponding codes for each theme are listed below.

Table 8.2 explains pre-service teachers' general perceptions of AI tools and their awareness of the contributions these tools make to the teaching process at the elementary school level.

Table 8.2, under the theme "Perception and Awareness," reflects how pre-service teachers perceive AI tools at the primary school level and their awareness of them. In the "Contribution to the Learning Process" category, participants emphasized that using AI tools makes learning more fun and engaging; they noted that elements such as music, visual content, and gamification increase student engagement. In the permanence and attention category, they stated that ChatGPT and Kahoot effectively produce engaging and interactive content in classroom applications and that visual materials prepared with Canva-Al can increase student interest. In the Teacher Assistance category, they stated that they support teachers in lesson plan preparation, content development, and measurement-assessment processes. In the "Adaptation to Technology" category, they said that it is necessary for the digital era, making lessons more productive by keeping students' attention throughout the lesson, and facilitating the adaptation of students and teachers to technological tools. Some participant views on this matter are as follows:

Table 8.2 Perception and awareness

Pre-service Teachers' Perceptions and Awareness Levels Regarding AI Tools	
Category	Code
Contribution to The Learning Process	Fun learning, active participation, learning through music, and visual support
Permanence and Attention	Permanent learning, attracting attention, and preventing boredom
Teacher Assistance	Lesson plan support, content development, measurement-assessment, and adapting to the lesson
Adapting to Technology	Digital era necessity, dependence on technological tools

- *PT2: "Making learning more engaging draws attention when they're bored."*
- *PT5: "Lessons become more engaging with pictures and music."*
- *PT8: "It will be beneficial for teachers when preparing lesson plans. In the era of technology, it attracts students' attention and makes the lesson more productive."*
- *PT11: "They are surrounded by technological devices at home and outside, and to attract their attention, we need to continue learning at school.*

2. Table 8.3 explains pre-service teachers' views on AI tools in relation to teacher roles and responsibilities.

Table 8.3 addresses how AI tools may affect teachers' job descriptions under the "Teacher Roles and Responsibilities." Participants in the role change category stated that teachers' duties and responsibilities as guides may increase, and that the definition of teaching must evolve to suit the new era. In the "facilitator function" category, they indicated that AI could reduce teachers' workload, but that excessive dependence could limit pedagogical creativity.

In the "risks and dependency" category, it was noted that potential risks could create dependence on AI, and that overuse of AI tools could distract students' attention and interest, leading to limited idea generation. At the same time, it was stated that this entire communication process strengthens the emotional bond between teachers and students. Some views are as follows:

- *PT4: "It makes things easier for teachers and can enrich the teachers themselves, but if the teachers become dependent on AI, they may become disconnected from the classroom, and new ideas will not emerge."*
- *PT7: "It reduces the teacher's workload and allows them to interact more with students."*
- *PT11: "Teachers can renew themselves due to preparedness. However, they can also adapt and integrate into today's world."*

3. Table 8.4 explains pre-service teachers' awareness of the classroom usefulness and applicability of AI tools

Table 8.4 shows that participants indicated that using AI tools provided "Advantages" such as quick access within the classroom, learning efficiency, enabling personalized

Table 8.3 Teacher roles and responsibilities

The Impact of AI Tools on Teachers' Roles and Responsibilities	
Category	Code
Role Change	Guidance role, increased duties and responsibilities, transformation of the definition teaching
Facilitator Function	Reducing workload, facilitating the teaching process, gamificiation
Risks and Dependency	Dependency on AI, disconnection from the classroom, decrease of idea generation
Emotional Bond	Emotional communication

Table 8.4 Usability and applicability

The Usability and Applicability of AI Tools in the Classroom	
Category	Code
Advantages	Quick access, personalization, increased motivation, and permanent learning
Limitations	Access issues, rural disadvantages, membership requirment, getting used to convenience
Suitability for Lessons	Visual-auditory lessons, science, math, language teaching (Turkish-English)

teaching, and achieving permanent learning. However, issues such as access problems in rural areas and the requirement for membership for some applications were mentioned in relation to "Limitations."" The "Suitability for Lessons" category stated that it could be used in other subjects suitable for visual-auditory content, primarily in Science, Math, Turkish, and English lessons. Some views are as follows:

- *PT3: "It may be a disadvantage in rural village schools."*
- *PT10: "There should not be too much screen time; there should be learning activities in nature."*
- *PT5: "It increases participation in class because students learn more permanently through activities such as listening to music."*
- *PT7: "It makes learning permanent, makes the teacher's job easier, makes learning efficient, and because it prepares many things for us easily, it can make us complacent and lazy."*

4. Table 8.5 explains pre-service teachers' awareness of the integration of AI tools into pedagogical processes and their effects on students.

According to Table 8.5, participants indicated in the "Student Motivation" category that students can produce enjoyable content, particularly through visual and musical activities. In the "Creativity and Success" category, it was stated that thinking outside the box and gaining different perspectives can enhance creativity, and that such content can positively contribute to student success. The "Teaching Methods Framework" category stated that AI tools could be adapted to teaching methods by enriching

Table 8.5 Pedagogical integration and student impact

Integration of AI Tools into Pedagogical Processes and Their Effects on Students	
Category	Code
Student Motivation	Visual and musical attention-grabbing, fun contents
Creativity and Success	Increasing creativity, contributing to success, and thinking outside the box
Teaching Methods Framework	Quizzes, music, group work, game-based learning

them with game-based learning and group work, using music, and conducting quizzes for assessment. The participants' views on these are as follows:

- *PT1: "Student's motivation and current success will increase, creativityand participation in class will increase."*
- *PT8: "Unconventionally teaching the lesson increases the student's desire to learn."*
- *PT10: "Music and pictures attract attention; they also require compatibility with teaching methods."*
- *PT6: "Music, pictures, visual, and video-style content attract attention and are important for situations such as entertainment and motivation."*

5. Table 8.6 explains pre-service teachers' awareness of the potential risks that may arise in terms of ethics and security when using an AI tool.

Table 8.6 shows that participants indicated that AI tools may pose certain risks regarding personal data security and ethical values in the "Privacy and Security Risk" category, and that they may cause a decline in creativity by making individuals lazy in the "Ethical Risks" category. To mitigate these risks, as seen in the "Precautions" category, it is recommended to impose time limits when using these tools and to encourage students to use them by conducting research. Some views are as follows:

- *PT1: "It may compromise the security of personal information access."*
- *PT5: "If we provide ready-made information, they may get used to it after a while and stop researching."*
- *PT9: "It poses a risk in terms of security, may not be appropriate in terms of privacy, and ethical values may lead to corruption of ethical values in children."*
- *PT12: "Today, due to asociality, there are already issues such as cultural corruption and isolation that disregard ethics, and doubts about the security of the information provided reinforce each other."*

Table 8.7 explains pre-service teachers' readiness to use AI tools and their awareness of their professional development needs.

Table 8.7 indicates that for participants to use AI tools effectively, pre-service teachers must possess the "necessary knowledge and skills," and students must have digital literacy, inquiry skills, and technical and computer knowledge. Participants emphasized the importance of digital literacy and inquiry skills. In the "Need for Professional Support" category, it was stated that numerous courses should be offered on

Table 8.6 Ethical and security risks

Risks That May Arise in the Use of AI Tools in Terms of Ethics and Security	
Category	Code
Privacy and Security Risk	Personal data risk, data sharing, security vulnerabilities
Ethical Risks	Decreased creativity, cultural decay, laziness
Precautions	Time limits, data usage restrictions, research incentives

Table 8.7 Readiness and professional development

Pre-service Teachers' Readiness to Use AI Tools and Their Professional Development Needs	
Category	Code
Necessary Knowledge and Skills	Digital literacy, computer knowledge, inquiry skills
Need for Professional Support	Courses, university classes, collaboration with guidance counselors
Continous Learning	Lifelong learning, adapting to technology

these topics and that they should be included in university courses. Furthermore, it was explained that the use of AI tools should be integrated into lifelong learning, as seen in the "Continuous Learning" category, as an area appropriate for technology. Some views on this subject are as follows:

- *PT2: "Sufficient technological knowledge and conceptual understandingare needed."*
- *PT5: "Teachers should not repeat themselves but should be able to draw as much as they want from the ocean of knowledge."*
- *PT9: "Teachers should remain active and update themselves through lifelong learning."*

8.5 Discussion and Conclusion

When examining the results obtained from the research, pre-service teachers stated that they perceived the use of AI tools for teaching purposes as making the learning process enjoyable and engaging, thereby promoting retention, and that the use of music, visuals, and gamification would contribute to class participation, interest, motivation, and academic achievement. Kestin et al. (2025) stated that AI-based training would help students learn more quickly and be more interested and motivated than classroom learning. Pre-service teachers noted that using AI tools, such as game-based learning, group work, and assessment tools, is compatible with teaching methods. These tools can be used in other audio-visual content-based courses, particularly in science, mathematics, Turkish, and English. Casal-Otero et al. (2023)argued that AI literacy should be incorporated into the curriculum in a way suitable for an interdisciplinary and competency-based structure, stating that a new AI course is unnecessary. According to this, AI literacy can go beyond a single disciplinary concept by being based on the competencies and content of courses suitable for interdisciplinary work. In this regard, it can be interpreted that AI can be used effectively and actively in many interrelated project-based courses.

Another finding is that pre-service teachers believe that the definition of teaching needs to evolve to suit the times, and that in this context, teachers' duties and responsibilities may increase or, with the facilitating effect of AI, teachers' workload may be reduced. However, it was noted that excessive AI use could turn into an addiction, negatively affecting pedagogical creativity, which could also have adverse effects on students' attention and interest. It was also expressed that AI could strengthen the emotional bond between teachers and students when used effectively. In the study by Erdoğdu and Çakır (2024), pre-service teachers stated that AI in education has great potential. However, the existence of uncertainty and indecision in some areas is consistent with the findings of this study. The fact that pre-service teachers' perceptions of AI usage differ depending on the context suggests that the positive and negative aspects of AI usage in educational settings must be considered together. This is because teachers are the ones who will ensure the effective use of AI in classrooms, and they are the ones who decide when and how to apply these tools (İşler & Kılıç, 2021). Zhang et al. (2023) stated that more information is needed about pre-service teachers' acceptance of AI because their attitudes toward AI technologies significantly impact future students.

With the introduction of AI applications into the classroom, teachers are now expected to possess different skills (İşler & Kılıç, 2021). In this study, pre-service teachers expressed the knowledge and skills teachers need to use AI tools effectively. Accordingly, pre-service teachers are expected to possess digital literacy, computer skills, and inquiry skills in line with the principle of lifelong learning and adaptation to technology. A review of the relevant literature reveals that studies indicate that for teachers to use AI tools effectively, they need digital skills, critical thinking, operational skills, problem-solving, application skills, lifelong learning, ethical awareness, communication, collaboration/teamwork, self-regulation, creativity/innovation, emotional intelligence, reflection, adaptability, and flexibility. (AI Pioneers Project, 2024; Polak et al., 2022; Salas-Pilco et al., 2022; UNESCO, 2024). The need to offer courses to develop the skills that pre-service teachers will use in the process, to include AI-based courses in universities, and to collaborate with school guidance counselors has also been highlighted in the participants' views. It is essential to create detailed AI training programs for teachers and to include content in these trainings that will enable teachers to use AI technologies effectively. Furthermore, it is believed that a guide to be developed for teachers to use AI tools will provide significant support to teachers in the learning-teaching processes (MoNE IET, 2024).

Another finding of the study examined the usability and suitability of AI tools. According to this, the advantages of AI tools include providing quick access within the classroom, offering learning efficiency, enabling personalized teaching, and ensuring retention. The disadvantages include access problems in rural areas, and some applications require membership. Pre-service teachers also expressed concerns that privacy could be violated for reasons such as personal data risk, data sharing, and security vulnerabilities; they also expressed concerns that using AI tools could lead to decreased creativity, cultural degradation, and laziness. It was stated that time and data usage restrictions and research incentives could effectively prevent these ethical and security risks. Akgun and Greenhow (2022) noted that using AI in education carries potential risks such as privacy, bias and discrimination, surveillance, and autonomy. A review of the relevant literature reveals similar views to those of pre-service teachers. Understanding the positive features AI can offer education will make its use more effective, while knowing its risks will motivate us to use it in a controlled manner. Harry (2023) positively highlights AI's ability to personalize learning, increase lesson efficiency, improve student participation, save time, and provide consistent feedback; however, he also identifies privacy and security concerns, lack of trust, cost, and possible biases as challenges encountered in the use of AI tools. Despite all its challenges, using AI tools for teaching can revolutionize education integration and make learning and teaching processes personalized and efficient (Apolzan & Cimpieanu, 2024).

The study's findings reveal how pre-service teachers' experience AI tools in instructional design processes and how these experiences reflect on their pedagogical awareness and perceptions of professional development. The future generations will be raised by the future pre-service teachers in this era, the AI era. It is recommended that relevant individuals, institutions, and organizations integrate all the positive and negative features of AI use into education in a balanced manner and that teacher training programs be developed in line with the views of pre-service teachers.

References

AI Index Steering Committee. (2024). *AI Index report 2024*. Stanford University, Human-Centered Artificial Intelligence (HAI). https://arxiv.org/abs/2405.19522

AI Index Steering Committee. (2025). *AI Index report 2025*. Stanford University, Human-Centered Artificial Intelligence (HAI). https://arxiv.org/pdf/2504.07139

AI Pioneers Project. (2024). *Supplement to the DigCompEdu framework (WP3): AI competencies for educators* (AI Pioneers). https://aipioneers.org/wp-content/uploads/2024/01/WP3_Supplement_to_the_DigCompEDU_English.pdf

Akgun, S., & Greenhow, C. (2022). Artificial intelligence in education: Addressing ethical challenges in K-12 settings. *AI Ethics, 2*, 431–440. https://doi.org/10.1007/s43681-021-00096-7

Alejandro, I. M. V., Sanchez, J. M. P., Sumalinog, G. G., Mananay, J. A., Goles, C. E., & Fernandez, C. B. (2024). Pre-service teachers' technology acceptance of artificial intelligence (AI) applications in education. *STEM Education, 4*(4), 445–465. https://doi.org/10.3934/steme.2024024

Apolzan, I., & Cimpineanu, M. J. (2024). Benefits and challenges of using artificial intelligence in education. *Euro-Atlantic Resilience Journal, 2*(3), 49–72.

Ayanwale, M. A., Adelana, O. P., Molefi, R. R., Adeeko, O., & Ishola, A. M. (2024). Examining artificial intelligence literacy among pre-service teachers for future classrooms. *Computers and education open, 6*. https://doi.org/10.1016/j.caeo.2024.100179

Casal-Otero, L., Catala, A., Fernández-Morante, C., Taboada, M., Cebreiro, B., & Barro, S. (2023). AI literacy in K-12: A systematic literature review. *International Journal of STEM Education, 10*(1), 29. https://doi.org/10.1186/s40594-023-00418-7

Chen, L., Chen, P., & Lin, Z. (2020). Artificial intelligence in education: A review. *IEEE Access, 8*, 75264–75278. https://doi.org/10.1109/ACCESS.2020.2988510

Erdoğdu, F., & Cakir, O. (2024). Determining pre-service teachers' artificial intelligence literacy and perceptions of artificial intelligence. *Journal of Social Sciences in the International Turkish Cultural Geography, 9*(2), 63–95. https://doi.org/10.55107/turksosbilder.1594635

Harry, A. (2023). Role of AI in education. *Interdiciplinary Journal & Hummanity (INJURITY), 2*(3). https://doi.org/10.58631/injurity.v2i3.52

Holmes, W., Bialik, M. & Fadel, C. (2019). Artificial intelligence in education: Promises and implications for teaching and learning. *Center for Curriculum Redesign.* https://www.curriculumredesign.org/wp-content/uploads/AIED-Book-ExcerptCCR.pd
İşler, B., & Kılıç, M. (2021). The use and development of artificial intelligence in education. *New Media Electronic Journal, 5*(1), 1–11. https://doi.org/10.17932/IAU.EJNM.25480200.2021/ejnm_v5i1001
Kaya, M., & Köseoğlu, Z. (2024). Shaping the future of education: Artificial intelligence teaching assistant. *Pearson Journal, 8*(29), 1555–1578. https://doi.org/10.5281/zenodo.13384238
Kestin, G., Miller, K., Klales, A., Milbourne, T., & Ponti, G. (2025). AI tutoring outperforms in-class active learning: An RCT introducing a novel research-based design in an authentic educational setting. *Scientific Reports, 15*(1), 17458. https://doi.org/10.1038/s41598-025-97652-6
Kottara, C. (2025). Application of artificial intelligence. In education: Potentialities and dilemmas. *European Journal of Education Studies, 12*(11). https://doi.org/10.46827/ejes.v12i11.6295
Long, T., & Johnson, M. (2000). Rigour, reliability and validity in qualitative research. *Clinical Effectiveness in Nursing, 4*(1), 30–37. https://doi.org/10.1054/cein.2000.0106
Lopez, K. A., & Willis, D. G. (2004). Descriptive versus interpretive phenomenology: Their contributions to nursing knowledge. *Qualitative health research, 14*(5), 726–735.https://doi.org/10.1177/1049732304263638
Merriam, S. B. (2013). *Qualitative research: A guide to design and practice* (Trans. Ed. S. Turan). Nobel. (Original work published 2009).
Miles, M. B., & Huberman, A. M. (1994). *Qualitative data analysis.* Sage Publications.
Moemeke, C. D. (2024). Artificial intelligence and machine learning in enhancing science learning experiences: Exploring possibilities and concerns. *NIU Journal of Educational Research, 10*(2), 59–72.
Molina, E., Cobo, C., Pineda, J., & Rovner, H. (2024). *AI revolution in education: What you need to know.* In Digital Innovations in Education. World Bank.
MoNE IET (2024). *Report of the international forum on artificial intelligence applications in education.*https://yegitek.meb.gov.tr/meb_iys_dosyalar/2024_09/11104346_meb_egitimde_uyz_formu_raporu_web_28082024_tr.pdf
OECD (2025), *"What should teachers teach and students learn in a future of powerful AI?", OECD Education Spotlights*, No. 20, OECD Publishing. https://doi.org/10.1787/ca56c7d6-en.
Özdere, M. (2024). The future of the teaching profession in the age of artificial intelligence. In M. Alanoğlu (Ed.), *Education & Science 2024-III* (1st ed., pp. 31–50). Efe Akademi. https://doi.org/10.59617/efepub2024158
Petrone, J. (2025, March 19). AI Leap 2025: Estonia sets the global standard for AI in education. *E-Estonia.* https://e-estonia.com/ai-leap-2025-estonia-sets-ai-standard-in-education/
Polak, S., Schiavo, G., & Zancanaro, M. (2022, April). Teachers' perspective on artificial intelligence education: An initial investigation. In *CHI conference on human factors in computing systems extended abstracts* (pp. 1–7). https://doi.org/10.1145/3491101.3519866
Rahman, M. M., & Watanobe, Y. (2023). ChatGPT for education and research: Opportunities, threats, and strategies. *Applied Sciences, 13*(9), 5783. https://doi.org/10.3390/app13095783
Ryu, M., & Han, S. (2018). The educational perception on artificial intelligence by elementary school teachers. *Journal of The Korean Association of Information Education, 22*(3), 317–324. https://doi.org/10.14352/jkaie.2018.22.3.317
Saban, A. & Ersoy, A. (2019). *Qualitative research designs in education.* Ankara: Anı
Sadiku, M. N. O., Ashaolu, T. J., Ajayi-Majebi, A., & Musa, S. M. (2021). Artificial intelligence in education. *International Journal of Scientific Advances, 2*(1). https://doi.org/10.51542/ijscia.v2i1.2
Salas-Pilco, S. Z., Xiao, K., & Hu, X. (2022). Artificial intelligence and learning analytics in teacher education: A systematic review. *Education Sciences, 12*(8), 569. https://doi.org/10.3390/educsci12080569
Sanusi, I. T., Ayanwale, M. A., & Tolorunleke, A. E. (2024). Investigating pre-service teachers' artificial intelligence perception from the perspective of planned behavior theory. *Computers*

and Education: Artificial Intelligence, 6, Article 100202. https://doi.org/10.1016/j.caeai.2024.100202

Trilling, B., & Fadel, C. (2009). *21st century skills: Learning for life in our times*. Jossey-Bass.

U.S. Department of Education, Office of Educational Technology, (2023). *Artificial intelligence and future of teaching and learning: Insights and recommendations*. https://www.ed.gov/sites/ed/files/documents/ai-report/ai-report.pdf

UNESCO (2024). *AI competency framework for teachers*. https://doi.org/10.54675/ZJTE2084

Wang, X., Li, L., Tan, S. C., Yang, L., & Lei, J. (2023). Preparing for AI-enhanced education: Conceptualizing and empirically examining teachers' AI readiness. *Computers in Human Behavior, 146*. https://doi.org/10.1016/j.chb.2023.107798

Yao, N., & Wang, Q. (2024). Factors influencing pre-service special education teachers' intention toward AI in education: Digital literacy, teacher self-efficacy, perceived ease of use, and perceived usefulness. *Heliyon, 10*(14). https://doi.org/10.1016/j.heliyon.2024.e34894

Yazıcı Demir, Y. (2025). 21st-century competencies and the new generation teacher profile. In A. Kızılkaya (Ed.), *Conceptual foundations of education-12-digitalization and new pedagogies* (1st ed., pp. 23–40). Efe Akademi. https://doi.org/10.59617/efepub2025122

Yue, M., Jong, M. S. Y., & Dai, Y. (2022). Pedagogical design of K-12 artificial intelligence education: A systematic review. *Sustainability, 14*(23), 15620. https://doi.org/10.3390/su142315620

Yue, M., Jong, M. S. Y., & Ng, D. T. K. (2024). Understanding K–12 teachers' technological pedagogical content knowledge readiness and attitudes toward artificial intelligence education. *Education and Information Technologies, 29*(15), 19505–19536. https://doi.org/10.1007/s10639-024-12621-2

Zhang, C., Schießl, J., Plößl, L., Hofmann, F., & Gläser-Zikuda, M. (2023). Acceptance of artificial intelligence among pre-service teachers: A multigroup analysis. *International Journal of Educational Technology in Higher Education, 20*(49). https://doi.org/10.1186/s41239-023-00420-7

Chapter 9
Generative Artificial Intelligence and the Development of Higher-Order Thinking Skills in K-12 Science Education: An Integrative Literature Review (2020–2025)

Rosa Neto, Isabel Saúde, and José Luís Araújo

9.1 Introduction

The emergence of Generative Artificial Intelligence (GenAI), accelerated by the release of ChatGPT in late 2022, has profoundly reshaped digital education (UNESCO, 2024a). Capable of generating text and images and supporting reasoning, GenAI shows unprecedented potential to emulate higher-order thinking, compelling institutions to reconsider the foundations of learning (UNESCO, 2023). Large Language Models (LLMs) demonstrate literacy, reasoning, and problem-solving capacities—especially in mathematics and science—comparable to, or exceeding,

The original version of the chapter has been revised. A correction to this chapter can be found at https://doi.org/10.1007/978-3-032-18030-8_14

R. Neto · J. L. Araújo (✉)
Paços de Ferreira Secondary School, Paços de Ferreira, Portugal
e-mail: jlaraujo@ua.pt

R. Neto
e-mail: rmarianeto@ua.pt

R. Neto
Department of Education and Psychology, University of Aveiro, Aveiro, Portugal

I. Saúde
Faculty of Sciences, Science Teaching Unit, University of Porto, Porto, Portugal
e-mail: isabel.saude@ua.pt

J. L. Araújo
CIDTFF, Department of Education and Psychology, University of Aveiro, Aveiro, Portugal

© The Author(s), under exclusive license to Springer Nature Switzerland AG 2026, corrected publication 2026

M. Alkış Küçükaydın (ed.), *Preparing K-12 Pre-Service Science Teachers for the Age of Artificial Intelligence*, Advances in Technology-Rich Science Education 4,
https://doi.org/10.1007/978-3-032-18030-8_9

those of 15-year-old students in international assessments (OECD, 2023). These developments underscore the urgency of fostering Higher-Order Thinking Skills (HOTS) (UNESCO, 2023).

HOTS involve analysis, evaluation, and creation (Facione, 1990), supported by metacognitive and affective dispositions (Facione, 1990; Paul & Elder, 2018), and are central to STEM learning. GenAI's dual nature—its cognitive potential and risk of superficiality—has been described as a "double-edged sword" (Kasneci et al., 2023). Empirical evidence shows that positive outcomes depend on pedagogical orchestration, particularly when GenAI is embedded in active learning, adaptive tutoring, co-creation, or simulation approaches; these effects weaken when teacher mediation is limited (Deng et al., 2025).

When used as an active cognitive mediator or "Socratic opponent," GenAI fosters critical thinking and AI literacy by prompting students to evaluate and refine its outputs (UNESCO, 2025). The UNESCO Competency Frameworks for Students (MRCE-IA) and Teachers (MRCP-IA) (2024a, 2024b) offer conceptual and pedagogical guidance for ethical and human-centered integration.

Despite the need to equip K–12 learners in science and mathematics with such competencies, research remains concentrated in higher education (Deng et al., 2025; Dimeli & Kostas, 2025). Broader reviews on AI in science education (Almasri, 2024) and STEM in secondary education (Akhmetova et al., 2025) confirm the absence of syntheses addressing GenAI with HOTS as a primary outcome. This gap underscores the need for studies linking empirical evidence with pedagogical and contextual dimensions.

Given the heterogeneity of designs and metrics, an integrative literature review is the most suitable approach (Whittemore & Knafl, 2005), enabling the synthesis of diverse theoretical and empirical sources to inform both research and practice. This review aims to fill that gap by analyzing GenAI's role in fostering HOTS in K–12 science, mathematics, and STEM. It addresses the central research question:

> How does the use of generative artificial intelligence (GenAI) contribute to the development of higher-order thinking skills in non-tertiary students, particularly in science and mathematics?

To ensure depth and transparency, the review is further guided by four secondary research questions (QI):

- QI1. What educational practices and pedagogical strategies using GenAI are documented as fostering HOTS in K–12 science and mathematics?
- QI2. What impacts, limitations, and challenges—cognitive, ethical, and evaluative—are reported in the literature regarding the integration of GenAI into HOTS development?
- QI3. What perceptions and contextual factors influence the use of GenAI in K–12 science and mathematics education?
- QI4. What theoretical models or conceptual contributions (e.g., competency frameworks) exist to frame the integration of GenAI in the development of HOTS in K–12 contexts?

9.2 Background

Understanding the pedagogical potential of Generative Artificial Intelligence (GenAI) for developing higher-order thinking skills (HOTS) requires situating it within its conceptual and educational context. While the complexity of bridging human and machine intelligence long constrained progress, large language models (LLMs) now generate multimodal outputs with human-like fluency (Feuerriegel et al., 2023).

The mimetic capacity of GenAI has inspired divergent interpretations—from a "Pandora's box" reshaping civilization (Harari, 2018) to a co-intelligent collaborator extending human capacities (Mollick, 2024). A hybrid view emphasizes *sympoiesis*, or "making-with" human and non-human intelligences (Haraway, 2016). ChatGPT's rise exemplified this shift, enabling applications such as content generation, inquiry stimulation, and personalized learning (Hsu and Ching 2023), while raising concerns about overreliance and reliability (Mollick & Mollick, 2023; UNESCO, 2023).

The theoretical basis of HOTS lies in critical and creative thinking. Rooted in Dewey's "reflective thinking," critical thinking entails self-regulated analysis, inference, and evaluation, supported by curiosity and intellectual humility (Ennis, 2015; Facione, 1990). Creative thinking, as imagination and innovation—or "lateral thinking" (De Bono, 1995)—complements critical reasoning (Lipman, 1995; Paul & Elder, 2018). Both constitute HOTS' core and align with Bloom's revised taxonomy, where analysis, evaluation, and creation occupy the highest cognitive levels, supported by metacognition for autonomy and transfer (Anderson & Krathwohl, 2001).

Problem solving and computational thinking are major HOTS manifestations. They involve analysis, creativity, and metacognition in complex, open-ended contexts. Problem solving applies critical reasoning through hypothesis formulation, evaluation, and decision-making (Ennis, 2015), while computational thinking emphasizes decomposition, abstraction, and algorithm design as key STEM competencies (Grover & Pea, 2013).

Developing these skills requires process-oriented pedagogy and assessment—reasoning traceability, rubrics, iterative review—to mitigate bias and hallucinations (UNESCO, 2023). In K–12 science and mathematics, this includes hypothesis formulation, modelling, data analysis, and evidence evaluation (NGSS Lead States, 2013). Global frameworks underscore HOTS, metacognition, and AI literacy as essential for citizenship in the digital age (OECD, 2021; UNESCO, 2021). With models like GPT-4 already surpassing 15-year-olds in reading and science (OECD, 2023), schools must emphasize critical oversight, creativity, and ethical use.

Recent advances—multimodality, assisted reasoning, and human-supervised agents—highlight the need for visible reasoning through planning, hypothesis testing, and verification (UNESCO, 2024a). UNESCO frames AI as a public good, requiring equitable and accountable integration. Its AI literacy frameworks define student competencies (human-centered mindset, ethics, design) and teacher roles in AI pedagogy and epistemic curation, positioning HOTS as the culminating goal (UNESCO, 2024a, 2024b).

Overall, GenAI acts as a powerful cognitive mediator with dual potential: when critically integrated, it enhances problem solving and decision-making; when used passively, it risks cognitive offloading and reduced autonomy (Stadler et al., 2024). Despite expanding research, most reviews focus on higher education, leaving a gap in K–12 science and mathematics—justifying the present integrative review.

9.3 Main Focus of the Chapter

9.3.1 Methodology of the Review

This review addressed the central question of how Generative Artificial Intelligence (GenAI) contributes to the development of higher-order thinking skills (HOTS) in non-tertiary science education. An integrative literature review approach was adopted (Whittemore & Knafl, 2005), which combines empirical and conceptual works across experimental and non-experimental designs, offering a holistic perspective on an emerging phenomenon. To ensure rigor and transparency, the process followed PRISMA guidelines.

The search, conducted in July–August 2025, covered three major databases—Scopus, Web of Science (WoS), and ERIC—chosen for their complementary coverage. The time frame was restricted to 2020–2025, corresponding to the rapid growth of GenAI research, especially after the public release of ChatGPT in late 2022. Only peer-reviewed journal articles in English, Portuguese, or Spanish were considered, although all retrieved works were in English.

The search strategy was structured into four conceptual blocks—Generative AI technologies, Higher-order thinking skills (HOTS), scientific disciplines, and Non-tertiary education (K–12)—combined through Boolean operators. The descriptors are summarized in Table 9.1.

Eligibility criteria included studies that: (i) explicitly examined GenAI use; (ii) addressed HOTS such as critical thinking, creativity, problem solving, computational thinking, metacognition, or self-regulation; (iii) focused on K–12 science or mathematics contexts (including STEM with a scientific component); and (iv) provided empirical or conceptual insights. Exclusion criteria eliminated works on non-generative AI, non-HOTS outcomes, or higher/professional education only.

The process identified 139 records, reduced to 99 after duplicates. After title/abstract screening, 27 articles were retained for full-text review; three were inaccessible and six excluded, leaving 18 studies in the final synthesis. The PRISMA flow diagram summarizing identification, screening, and inclusion is shown in Fig. 9.1.

Table 9.1 Conceptual blocks and descriptors used in the search

Conceptual Block	Descriptors
Generative AI technologies	• ChatGPT; ChatBot; GPT; "generative AI"; GAI; GenAI; "gen AI"; "generative artificial intelligence"; "large language models"; "LLM"; "artificial intelligence".
Higher-order thinking skills (HOTS)	• "higher-order thinking"; "higher order thinking"; "critical thinking"; "creative thinking"; "problem solving"; metacognition; "computational thinking"; "self-regulation"; "self regulation".
Scientific disciplines	• "science education"; "physics education"; "chemistry education"; "earth science education"; "astronomy education"; "STEM education"; "environmental science education"; "biology education"; mathematics.
Non-tertiary education (K–12)	• "secondary education"; K-12; "primary school"; "secondary school"; "middle school"; "high school"; "elementary education"; "basic education"; "compulsory education"; "pre-university education".

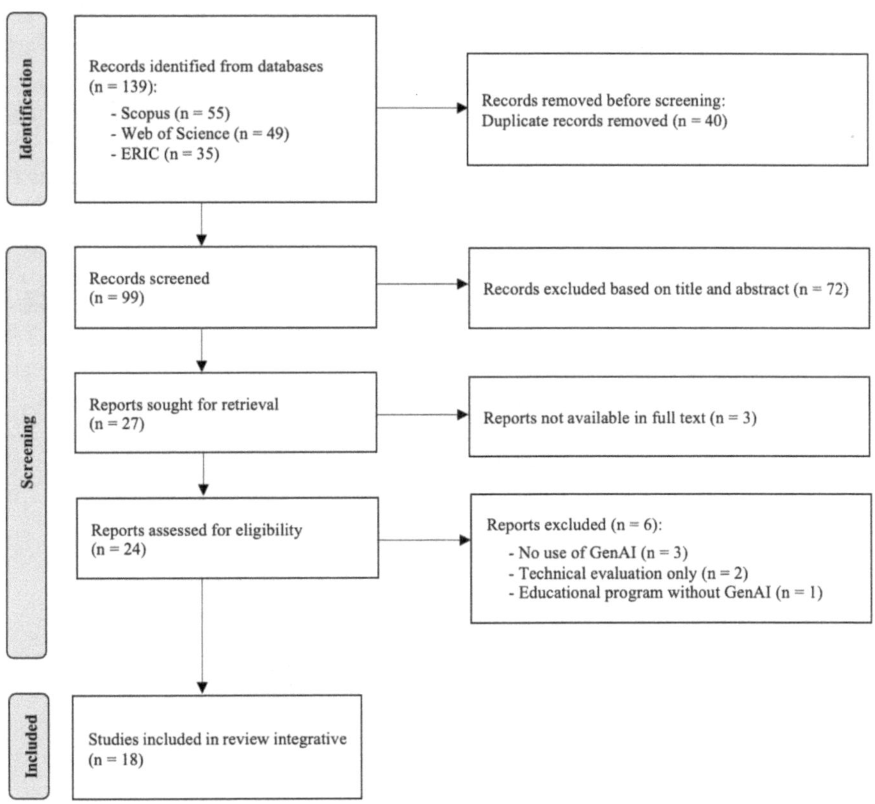

Fig. 9.1 PRISMA flow diagram adapted for the integrative review

Table 9.2 Study distribution by analytical category

Category	Description	Authors
A	Interventions with direct measurement of HOTS.	Guo et al. (2025); Lin et al. (2025); Tong et al. (2025); Liu et al. (2025); Xuan et al. (2025); Baroud et al. (2025).
B	Interventions with indirect assessment of HOTS (proxies).	Bitzenbauer (2023); Fang et al. (2025); Baez et al. (2025); Yunianto et al. (2024); Tang and Putra (2025).
C	Perception studies (teachers or students, without structured intervention).	Adeyele and Ramnarain (2024); Jang and Choi (2025); Fadillah et al. (2024); Al-Karasneh et al. (2025).
D	Conceptual/theoretical contributions.	Sirisathitkul et al. (2025); Kunnath and Botes (2025); Chau et al. (2025).

9.3.2 Analysis, Synthesis and Categorization of Studies

The review identified 18 studies published between 2023 and 2025 that address the role of Generative Artificial Intelligence (GenAI) in fostering higher-order thinking skills (HOTS) in K–12 science and mathematics education. The field is emerging, geographically diverse, and methodologically heterogeneous, ranging from quasi-experimental interventions to case studies, perception surveys, and conceptual analyses. Findings converge that GenAI supports HOTS only when orchestrated as a cognitive mediator within inquiry-oriented pedagogies, with effectiveness shaped by design, teacher mediation, and contextual factors.

For comparative purposes, studies were grouped into four categories: (A) classroom interventions with *direct* HOTS measures; (B) interventions inferring HOTS from proxies; (C) *teachers' and students' perception studies*; and (D) *conceptual contributions.* Table 9.2 summarizes the distribution of studies across categories, highlighting their main focus, mechanisms of GenAI use, and outcome tendencies.

9.3.3 Category A—Interventions with Direct Measurement of HOTS

This category comprises six quasi-experimental classroom studies that directly assessed higher-order thinking skills (HOTS) in Mathematics, Science, or STEM contexts (Baroud et al., 2025; Guo et al., 2025; Lin et al., 2025; Liu et al., 2025; Tong et al., 2025; Xuan et al., 2025) (see Table 9.3). Two pedagogical patterns stand out. First, error-based dialogic mediation, where students analyzed and corrected GPT-4 outputs, proved effective in cultivating critical thinking and metacognition, particularly for lower-performing students (Lin et al., 2025). Second, adaptive scaffolding in complex tasks—from Science and Engineering Practices (Guo et al., 2025) to personalized mathematical problem-solving (Liu et al., 2025; Xuan et al., 2025) and genetics inquiry (Baroud et al., 2025)—enhanced computational thinking, problem-solving, and higher-order reasoning aligned with Bloom's taxonomy.

Three moderators emerged across studies:

- Duration—short, intensive sessions fostered immediate problem-solving gains (Liu et al., 2025), while multi-week interventions consolidated metacognitive growth (Guo et al., 2025; Lin et al., 2025).
- Interaction quality—when AI was reduced to a Q&A tool, learning was superficial, whereas dialogic or collaborative use promoted scientific reasoning (Tong et al., 2025).
- Context—resource disparities mattered: AI-assisted instruction benefited private school settings but was less effective in under-resourced public schools (Xuan et al., 2025).

Overall, Category A shows that GenAI's value depends less on technical sophistication than on intentional orchestration by teachers.

9.3.4 Category B—Interventions with Indirect Assessment of HOTS

This category includes five studies in which GenAI's impact on higher-order thinking skills (HOTS) was inferred from indirect measures such as perceptions, chatlogs, learning behavior patterns, or external ratings (Baez et al., 2025; Bitzenbauer, 2023; Fang et al., 2025; Yunianto et al., 2024; Tang & Putra, 2025) (see Table 9.4).

Three mechanisms consistently emerge:

1. Dialogic interaction and Socratic prompting—DST chatbots stimulated reasoning, argumentation, and creativity (Tang & Putra, 2025), while exposing students to flawed AI outputs encouraged critical scrutiny, though effects were more attitudinal than cognitive (Bitzenbauer, 2023).
2. Generative feedback cycles—iterative feedback on geometry reflections (Gemini) improved the depth and lexical richness of student writing, suggesting plausible gains in creativity and critical thinking (Baez et al., 2025).
3. AI-based scaffolding—pedagogical agents offering personalized, multimodal support fostered collaboration, problem-solving, and self-regulation (Fang et al., 2025). Similarly, computational thinking tasks with ChatGPT revealed that outcomes depended strongly on prompt literacy, with vague prompts leading to failure (Yunianto et al., 2024).

Overall, Category B highlights process-oriented pathways—dialogicity, iterative feedback, and adaptive scaffolding—through which GenAI can shape learning. However, reliance on proxies makes evidence less conclusive, and outcomes depend strongly on prompt literacy.

Table 9.3 Category A—Interventions with Direct Measurement of HOTS: Study Characteristics and Pedagogical Approaches

Author/ Year	Sample & Design	Intervention (GenAI Mechanism)	HOTS Measured (Direct Measurement)
Guo et al. (2025)	N = 70, grade 8 (China); quasi-experimental (LLMs vs. traditional SEP; 4 weeks/4 lessons + pre-class briefing).	GLM-4 integrated into science and engineering practice (submarine project); Socratic guidance, modeling, coding/ debugging; teacher-mediated scaffolding.	Computational thinking (abstraction, decomposition, algorithmic reasoning); STEM problem solving; scientific knowledge (validated instruments).
Liu et al. (2025)	N = 104, grade 5, China); quasi-experimental (ChatGPT-MPS vs. conventional; 10-min briefing +40-min session).	ChatGPT-MPS (GPT-4/4o API): Personalised scaffolding, step-by-step guidance, multiple solution pathways for Word problems.	Mathematical problem-solving (parallel pre/post tests aligned to curriculum). *(SISPS = affective, not HOTS.)*
Lin et al. (2025)	N = 59, grade 5, Taiwan); quasi-experimental (GPT-4 error analysis vs. traditional instruction); 10 weeks, 2 sessions/ week).	Collaborative error analysis of GPT-4-generated solutions to authentic mathematical problems (correct/incorrect/ incomplete); peer explanation.	Authentic mathematical problem solving (AMPS test). *Note:* Mathematical confidence is affective, not HOTS.
Xuan et al. (2025)	N = 120, high school (Vietnam); quasi-experimental (ChatGPT 3.5, AI-assisted vs. teacher-supported across public/private comparison; 2 weeks).	ChatGPT 3.5-supported tailored explanations and practice on limits in algebra; contrast with teacher-centred support.	Higher-order cognitive performance (bloom-aligned higher-level items: Analysis/synthesis/ evaluation) + total achievement.
Tong et al. (2025)	N = 115, high school (China); quasi-experimental (human-AI collaboration vs. human-human collaboration; 4 sessions/ 130 min total).	Human–AI collaboration in open-ended physics problem-solving; ChatGPT-4o as collaborative partner.	Scientific thinking (control of variables, data analysis, causal reasoning) using iSTAR test and physics knowledge.
Baroud et al. (2025)	N = 111, second-year baccalaureate (Morocco); quasi-experimental (ChatGPT-supported genetics; 7 sessions).	ChatGPT as disciplinary resource to clarify complex genetics content (meiosis); guided in-class use.	Bloom levels: Knowledge, comprehension, analysis, application; N-gain and t-test (only analysis and application qualify as HOTS.)

Table 9.4 Category B—Interventions with Indirect Assessment of HOTS: Study Characteristics and Mechanism

Author/Year	Sample & Design	Intervention (GenAI Use)	HOTS Measured (Proxy)
Bitzenbauer (2023)	N = 53, grade 12 (Germany); questionnaire-based pilot study (2 physics lessons).	Physics activities fostering critical review of ChatGPT outputs and item writing.	Student perceptions of ChatGPT/AI usefulness and integration into daily life (questionnaire).
Tang and Putra (2025)	N = 21 (16–17 years), Indonesia; instrumental case study (chatlog analysis).	Dialogic science teaching (DST) chatbot engaging in Socratic questioning and reverse prompting on science topics.	Dialogic characteristics (perspective-taking, reasoning, argumentation, creative thinking) quantified from chatlogs.
Fang et al. (2025)	N = 47 (13–14 years), China; quasi-experimental (control vs. AIPA-STEAM).	STEAM course supported by an AI pedagogical agent (AIPA) offering personalised scaffolding and multimodal feedback.	Problem-solving and collaboration tendencies (ANCOVA); learning behavior patterns (LSA).
Baez et al. (2025)	High school geometry, USA; case study (11 student groups; 31 teacher evaluators).	Revision of reflections and interpretations with Gemini feedback.	Quality of written interpretations (Para-1 vs Para-2 with AI); external ratings as proxies for creativity and critical thinking.
Yunianto et al. (2024)	N = 16 (13–14 years), Indonesia; design-based research.	Mathematical computational thinking tasks using ChatGPT for GeoGebra code generation and debugging.	Student opinions and experiences; success in command generation/ debugging (proxies for CT facets).

9.3.5 Category C—Perceptions of Teachers and Students

This category comprises four studies mapping how teachers and students perceive GenAI's role in fostering higher-order thinking skills (HOTS) in science and mathematics education (Adeyele & Ramnarain, 2024; Al-Karasneh et al., 2025; Fadillah et al., 2024; Jang & Choi, 2025) (see Table 9.5).

Teachers generally described GenAI as a complementary resource for lesson planning, inquiry facilitation, and personalized instruction (Adeyele & Ramnarain, 2024; Al-Karasneh et al., 2025; Jang & Choi, 2025).

From the student perspective, Fadillah et al. (2024) found that perceptions of Convenience and Quality (CQ) and Motivation and Engagement (ME) predicted

Table 9.5 Category C—Teachers' and Students' Perceptions of GenAI in Science and Mathematics Education

Author/Year	Sample & Method	Perceived Benefits / Potential (HOTS-related)	Perceived Risks / Concerns / Barriers
Adeyele and Ramnarain (2024)	N = 15 science Teachers (JSS/SSS, Nigeria); qualitative (focus groups).	Teacher support within the IBL 5E framework (lesson planning, question generation); personalised instruction; enhanced engagement and motivation; fosters critical thinking and problem-solving.	Variable variable awareness/ adoption; risk of superficial IBL enactment without critical intentionality; need for targeted professional development for IBL-GenAI integration.
Jang and Choi (2025)	N = 10 physics educators (high school/ university, Korea); qualitative (CQR, semi-structured interviews).	Supports personalized, learner-led IBL; functions as a high-level cognitive tool for problem solving and scientific reasoning; can seed task and problem creation.	Uncertain reliability("hallucination"); risk of over-reliance/reduced practical reasoning; query-dependence (prompt quality); concerns over teacher authority; equity if access/ prompting skills differ; need to enhance digital competencies.
Fadillah et al. (2024)	N = 334 upper-secondary Students (Indonesia); quantitative (multiple linear regression).	CQ (Convenience & Quality) and ME (Motivation & Engagement) significantly predict students' HOTS in physics inquiry, supporting critical thinking and deeper conceptual analysis.	Risk of over-reliance potentially reducing independent and deep thinking.
Al-Karasneh et al. (2025)	N = 43 primary science Teachers (Saudi Arabia); qualitative (semi-structured interviews).	Stimulates creativity, problem-solving, curiosity; simplifies complex concepts with interactive resources; supports personalized learning; encourages inquiry-oriented and exploratory practices.	Technology/access constraints; resistance to change; need for specialized training (implementation and frameworks); importance of prompting competencies.

learning in physics inquiry tasks. CQ was the strongest predictor, suggesting that students value the instructional quality of AI dialogue more than sheer accuracy.

Despite these positive views, risks were consistently noted: unreliability and hallucinations (Jang & Choi, 2025), over-reliance reducing independent reasoning (Fadillah et al., 2024), erosion of teacher authority, and equity issues linked to access and prompting competence (Al-Karasneh et al., 2025). Teachers themselves stressed the need for specialized training and ethical guidance to ensure critical and intentional integration (Adeyele & Ramnarain, 2024).

In summary, Category C presents a dual picture: GenAI is widely perceived as a promising tool to stimulate inquiry, creativity, and problem-solving, but sustainable integration requires teacher mediation, professional development, and equitable access to prevent risks from outweighing benefits.

9.3.6 Category D—Conceptual and Theoretical Contributions

This category includes three conceptual works that propose theoretical models for integrating GenAI into science and mathematics education (Chau et al., 2025; Kunnath & Botes, 2025; Sirisathitkul et al., 2025) (see Table 9.6). Despite lacking empirical validation, they provide frameworks that help explain why and under what conditions GenAI may support higher-order thinking skills (HOTS).

Three orientations emerge:

Table 9.6 Category D – Conceptual and Theoretical Contributions: Context, Mechanisms, and HOTS Targets

Author/Year	Focus / Context	GenAI Role / Mechanism	HOTS Targets
Kunnath and Botes (2025)	Science; inquiry-based learning (IBL/5E) under resource constraints.	Supports engage–explore–explain–elaborate–evaluate (personalization, virtual labs/simulations, formative prompts); teacher orchestration.	Analysis, evaluation, critical thinking, and problem solving—GenAI as an amplifier of IBL.
Sirisathitkul et al. (2025)	Science/ engineering; inquiry with everyday objects.	"Agent-to-think-with": Plan–interpret–verify during inquiry; cognitive scaffold.	Scientific reasoning (control of variables, evidence-based justification).
Chau et al. (2025)	Mathematics; personalized teaching.	Adaptive tutoring; real-time feedback; dynamic task adjustment; error management; prompt-generation skills.	Problem-solving competence; self-regulation; logical reasoning.

1. Inquiry-based learning (IBL/5E frameworks) (Kunnath & Botes, 2025).
2. GenAI as a "thinking partner." (Sirisathitkul et al., 2025)
3. Adaptive tutoring for self-regulation (Chau et al., 2025).

All three contributions emphasize that teacher orchestration is indispensable: educators must design authentic tasks, curate prompts, and set validation criteria so that GenAI functions as a cognitive mediator rather than a shortcut.

In sum, these conceptual models provide testable frameworks for positioning GenAI as a mediator of inquiry, reflection, and adaptive feedback. They reinforce the broader evidence base suggesting that pedagogical design and human mediation are decisive for harnessing GenAI's potential to cultivate HOTS in K–12 classrooms.

9.3.7 Cross-Analysis and Synthesis

Taken together, the 18 studies reviewed portray an emerging but rapidly maturing field of research on the role of Generative AI (GenAI) in fostering higher-order thinking skills (HOTS) in K–12 science and mathematics education. The corpus spans a continuum of evidence: from controlled classroom interventions with direct measurement of HOTS (Category A), to process-oriented proxies (Category B), to teacher and student perceptions (Category C), and finally to theoretical models that frame GenAI as a cognitive mediator (Category D). Despite methodological heterogeneity and common limitations—such as small samples, short durations, and reliance on self-reports or novelty effects (Deng et al., 2025; Dimeli & Kostas, 2025)—the body of evidence converges on a central finding: GenAI enhances HOTS only when intentionally orchestrated within pedagogical architectures that preserve human agency, critical reflection, and teacher mediation. This trajectory reflects the global shift from an instrumental view of AI to a more critical and human-centered paradigm, consistent with recent UNESCO (2023) and OECD (2023) guidelines that emphasize critical thinking, problem-solving, and digital literacy as core educational priorities in the age of AI.

9.4 Pedagogy, Human Mediation, and HOTS

Across categories, the evidence demonstrates that GenAI enhances higher-order thinking skills (HOTS) only when embedded in intentional pedagogical designs and mediated by teachers. Direct assessments (Category A) showed that error-based analysis (Lin et al., 2025) and adaptive scaffolding in disciplinary tasks (Baroud et al., 2025; Guo et al., 2025; Liu et al., 2025; Xuan et al., 2025) significantly improved critical thinking, metacognition, and problem-solving. By contrast, when AI was reduced to a Q&A tool, learning outcomes were shallow and limited to factual recall (Tong et al., 2025).

Teachers thus emerge as epistemic curators, framing authentic tasks, scaffolding reasoning, and validating outputs. This role aligns with conceptual models portraying GenAI as an "agent-to-think-with" (Sirisathitkul et al., 2025) or as a scaffold within IBL/5E cycles (Kunnath & Botes, 2025). Adeyele and Ramnarain (2024) further emphasize that only teachers with critical intentionality can mitigate risks and use GenAI to stimulate inquiry. International frameworks converge: the AI Competency Framework for Teachers (MRCP-IA) highlights gaps in training and stresses a human-centered mindset where teachers safeguard ethics, inclusion, and equity (UNESCO, 2024b).

Contextual factors also matter. Xuan et al. (2025) found AI-supported instruction effective in private schools but less so in under-resourced public settings. UNESCO (2023) warns that unequal access may exacerbate rather than reduce educational inequalities, while OECD (2023) underlines that AI's effectiveness depends more on human and contextual mediation than on technical sophistication. In short, pedagogy and context—not the tool itself—determine GenAI's value for HOTS.

9.5 The Duality of GenAI and Critical Literacy

The review highlights the dual nature of GenAI as both opportunity and risk. On the positive side, it can foster reasoning, creativity, and problem-solving (Jang & Choi, 2025; Tang & Putra, 2025). Yet concerns persist about hallucinations and unreliability (Mollick, 2024), over-reliance leading to diminished reasoning, and erosion of teacher authority (Jang & Choi, 2025). The risk of cognitive offloading and accumulation of "cognitive debt" is also noted (Gerlich, 2025).

To address this, scholars and institutions stress the need for critical AI literacy. Students must learn to interrogate outputs, validate reasoning, and recognize AI's epistemic limits (Sardi et al., 2025). UNESCO (2024b) and OECD (2023) similarly call for embedding evaluation, ethics, and proportionality into curricula. The evidence from Category B reinforces this: when students lacked prompt literacy, GenAI failed to support computational thinking (Yunianto et al., 2024). By contrast, structured exposure to flawed outputs encouraged critical reflection, albeit more attitudinal than cognitive (Bitzenbauer, 2023). Developing such literacy is essential if GenAI is to enhance HOTS rather than undermine them.

9.6 Methodological Limitations

Despite promising findings, the corpus has important weaknesses. Most studies relied on small samples and short interventions, limiting external validity and raising questions about sustainability. Effects may partly reflect novelty rather than durable learning (Deng et al., 2025; Dimeli & Kostas, 2025). Few examined transfer of skills across contexts or over time, and replication is scarce.

Another limitation is the reliance on indirect measures such as perceptions or chat-logs. While useful to reveal processes, these cannot substitute for direct performance-based assessments. Future work should adopt authentic, competency-based assessments (Deng et al., 2025) and longitudinal, multi-site designs to establish whether observed gains persist and generalize.

9.7 Synthesis

Taken together, the review suggests that GenAI's potential to enhance HOTS depends less on technical capacity and more on pedagogical design, teacher mediation, and contextual equity. When orchestrated intentionally, GenAI can foster analysis, evaluation, creativity, and metacognition (Guo et al., 2025; Lin et al., 2025; Tang & Putra, 2025). Without such orchestration, it risks reinforcing dependency, inequities, and superficial engagement. This conclusion converges with international policy frameworks positioning AI not as a replacement for human teaching, but as a catalyst whose value is mediated by human intentionality (OECD, 2023; UNESCO, 2023).

9.8 Solutions and Recommendations

The integration of Generative Artificial Intelligence (GenAI) into the development of Higher-Order Thinking Skills (HOTS) in K–12 education depends on intentional pedagogical design and sustained human mediation (Guo et al., 2025; Lin et al., 2025; Xuan et al., 2025). The teacher's role as a bridge and guide is decisive for transforming GenAI from an answer-obtaining tool into a cognitive mediator. At the same time, risks such as uncritical dependence, cognitive offloading, and digital disparity must be mitigated (Fadillah et al., 2024; Tong et al., 2025). Based on the reviewed evidence, recommendations can be articulated across three complementary levels—micro (pedagogical design), meso (institutional culture), and macro (policy and equity)—to orient practices and policies toward ethical and effective integration.

9.9 Micro Level—Pedagogical Design and Classroom Practices

For GenAI to foster HOTS, outputs should be treated as objects of scrutiny rather than final products. This can be achieved by:

- Implementing error analysis and metacognitive reflection: Designing tasks where students identify, justify, and correct AI-generated errors, thereby transforming its uncertain reliability into an opportunity for reasoning (Bitzenbauer, 2023; Jang & Choi, 2025; Lin et al., 2025).
- Integrating inquiry-based models (IBL/5E) with dialogic interaction: Positioning GenAI as a dialogic partner or "devil's advocate" to support argumentation, perspective-taking, and conceptual elaboration (Guo et al., 2025; Kunnath & Botes, 2025; Tang & Putra, 2025).
- Evaluating cognitive processes: Prioritizing traceability of reasoning and quality of justification over the correctness of answers, reducing technological dependency and rewarding reflective engagement (Chau et al., 2025).

9.10 Meso Level—Teacher Development and Institutional Culture

Schools need to provide conditions for ethical and competent adoption of GenAI by:

- Professional development with critical intentionality: Training teachers to orchestrate GenAI, exercise epistemic curation, and manage the uncertainty of large language models (LLMs) in real classrooms (Adeyele & Ramnarain, 2024; Guo et al., 2025).
- AI literacy and qualified prompting: Equipping both students and teachers to formulate rich prompts and critically verify AI outputs (hallucinations, bias, inaccuracy) (Bitzenbauer, 2023; Jang & Choi, 2025; Yunianto et al., 2024).
- Clear institutional protocols: Establishing policies regarding GenAI use in assessment, data protection, and HOTS rubrics, aligned with international frameworks (OECD, 2023; UNESCO, 2023).

Even with well-prepared teachers and internal guidelines, structural constraints may hinder effectiveness, requiring systemic measures.

9.11 Macro Level—Equity and Regulatory Frameworks

To ensure that GenAI does not widen inequalities, broader policies are needed:

- Digital equity and minimum infrastructure: Guaranteeing connectivity, devices, and access to quality LLMs. In resource-limited contexts, hybrid models with reinforced teacher mediation can compensate for technological gaps (Xuan et al., 2025).
- Responsible governance and regulation: Implementing standards for privacy, security, and bias mitigation consistent with the principles of the Council of Europe and UNESCO (Council of Europe, 2022; UNESCO, 2023).

In summary, GenAI's contribution to HOTS depends on multi-level alignment. At the classroom level, teachers should design tasks that turn AI fallibility into opportunities for reasoning. At the institutional level, training and protocols are essential to sustain intentional use. At the systemic level, equity and regulation must be guaranteed to ensure that GenAI serves as a catalyst for critical, creative, and reflective learning rather than as a source of dependency or exclusion.

9.12 Future Research Directions

Research on Generative Artificial Intelligence (GenAI) and HOTS in K–12 education must now move beyond exploration toward a sustained, interdisciplinary agenda. While existing studies are promising, they reveal recurring design limitations. Three priorities emerge.

First, methodological consolidation and sustainability. Longitudinal and comparative studies are needed to examine the long-term impact of GenAI on knowledge retention and self-regulation in learning (Lin et al., 2025). To strengthen external validity, future research should include large-scale, multi-site replications and the development of standardized, consistent instruments for measuring HOTS.

Second, human–AI interaction and critical literacy. Current evidence highlights a predominantly extractive use of GenAI, often reduced to an answer-obtaining tool (Tong et al., 2025; Yunianto et al., 2024). Further work should examine pedagogical models that foster metacognitive co-regulation, critical prompting, and systematic evaluation of AI outputs, given frequent inconsistencies (Jang & Choi, 2025; Tong et al., 2025). Tasks involving error analysis, where students interrogate and correct AI-generated content, appear particularly promising (Bitzenbauer, 2023).

Third, equity and contextual factors. GenAI's effectiveness remains highly dependent on context, with digital disparities limiting adoption in under-resourced schools (Xuan et al., 2025). Research should explore hybrid or offline models with intensive scaffolding to mitigate these barriers. Beyond STEM, future studies should also examine other curricular areas. Finally, the need for robust ethical oversight (Guo et al., 2025) suggests exploring specialized AI agents for monitoring responsible use.

Taken together, these directions—sustainability, literacy, and equity—outline a coherent agenda to transform GenAI from a novel intervention into a durable and inclusive pedagogical infrastructure.

9.13 Conclusion

This integrative review demonstrates that Generative Artificial Intelligence (GenAI) holds significant potential for developing Higher-Order Thinking Skills (HOTS)—particularly critical thinking, computational reasoning, and metacognition—in K–12 science and mathematics education. Its effectiveness, however, is not intrinsic to the technology but depends on intentional pedagogical design and strong human mediation.

The evidence shows that the uncertainty of GenAI outputs can be transformed into a valuable pedagogical resource when integrated into structured practices. Error-analysis tasks (Bitzenbauer, 2023; Lin et al., 2025) and inquiry-based approaches such as the 5E model or Science and Engineering Practices (SEP) (Guo et al., 2025; Kunnath & Botes, 2025) consistently foster critical thinking, creativity, and self-regulation, demonstrating how GenAI can act as a cognitive partner when embedded in dialogic and reflective learning environments (QI1).

At the same time, important limitations and challenges remain (QI2). GenAI use tends to be ineffective when it is passive or purely instrumental. By contrast, when teachers guide students in critical prompting, validation, and ethical literacy, the pedagogical value of GenAI increases (Fadillah et al., 2024; Yunianto et al., 2024). Nevertheless, persistent technical problems—such as hallucinations, inconsistencies, and bias—reinforce the need for manual verification and enhance digital capacity for both teachers and students.

Contextual and equity factors further shape the effectiveness of GenAI integration (QI3). Digital disparities remain a decisive barrier: in under-resourced public schools, traditional instruction sometimes produced more reliable outcomes than AI-supported teaching (Xuan et al., 2025). These findings highlight the need for hybrid approaches with intensive scaffolding and robust teacher training, reaffirming the teacher's role as ethical mediator and epistemic curator (Adeyele & Ramnarain, 2024; Guo et al., 2025).

Finally, conceptual contributions offer important guidance for the future (QI4). Models emphasizing metacognitive co-regulation (Lin et al., 2025) and human–AI collaboration (Tong et al., 2025) underline the conditions under which GenAI can sustain deep learning. Integrative frameworks such as SEP (Guo et al., 2025) and the agent-to-think-with model (Sirisathitkul et al., 2025) provide both ethical and operational foundations for positioning GenAI as a dialogic and cognitive infrastructure complementary to human teaching.

In sum, the pedagogical value of GenAI lies not in its automated answers but in its ability to stimulate questioning, reflection, and learner agency. As research is still at an exploratory stage (Guo et al., 2025), future studies should adopt longitudinal and multi-context designs, develop standardized HOTS assessment instruments—including for metacognition (Tong et al., 2025)—and ensure equitable access through low-bandwidth or offline solutions (Xuan et al., 2025). Ultimately, the irreplaceable role of the teacher as cognitive, ethical, and human mediator remains the cornerstone for ensuring that GenAI functions as a genuine amplifier and catalyst of deep learning (Adeyele & Ramnarain, 2024; Jang & Choi, 2025).

Funding This works was financially supported by National Funds through FCT – Fundação para a Ciência e a Tecnologia, I.P., under the CIDTFF R&D Unit, with the reference UID/00194/2025 (https://doi.org/10.54499/UID/00194/2025).

References

Adeyele, V. O., & Ramnarain, U. (2024). Exploring the integration of ChatGPT in inquiry-based learning: Teacher perspectives. *International Journal of Technology in Education, 7*(2), 200–217.

Akhmetova, A. I., Sovetkanova, D. M., Komekbayeva, L. K., Abdrakhmanov, A. E., Yessenuly, D., & Serikova, O. S. (2025). A systematic review of artificial intelligence in high school STEM education research. *Eurasia Journal of Mathematics, Science and Technology Education, 21*(4).

Al-Karasneh, S. M., Kanaan, E. M., Al-Barakat, A. A., AlAli, R. M., Zaher, A. M., & Ibrahim, N. A. (2025). Transforming primary science education: Unlocking the power of generative ai to enhance pupils' grasp of scientific concepts. *International Journal of Learning, Teaching and Educational Research, 24*(5), 304–322.

Almasri, F. (2024). Exploring the impact of artificial intelligence in teaching and learning of science: A systematic review of empirical research. *Research in Science Education, 54*(5), 977–997.

Anderson, L. W., & Krathwohl, D. R. (2001). *A taxonomy for learning, teaching, and assessing: A revision of bloom's taxonomy of educational objectives*. Longman.

Baez, R., Sanchez, H., Sandeep, N., & Pllana, D. (2025). Creativity and critical thinking in the AI age: A high school perspective. *International Journal of Research in Education and Science, 11*(2), 448–477.

Baroud, S., El Alaoui, M., Idrissi, R. J., Boucetta, N., Ghariz, G., & Seghir, H. (2025). Improving biology learning in high school with ChatGPT: A case study in Morocco. *Eurasia Journal of Mathematics, Science and Technology Education, 21*(7).

Bitzenbauer, P. (2023). ChatGPT in physics education: A pilot study on easy-to-implement activities. *Contemporary Educational Technology, 15*(3).

Chau, D. B., Luong, V. T., Long, T. T. & Linh, N. T. T. (2025). Personalized mathematics teaching with the support of AI Chabot's to improve mathematical problem-solving competence for high school students in Vietnam. *European Journal of Educational Research, 14*(1), 323–333.

Council of Europe. (2022). Artificial intelligence and education: A critical view through the lens of human rights, democracy and the rule of law. *Council of Europe.*. https://edoc.coe.int/en/artificial-intelligence/11269-artificial-intelligence-and-education-a-critical-view-through-the-lens-of-human-rights-democracy-and-the-rule-of-law.html

De Bono, E. (1995). *Parallel thinking: From Socratic thinking to De Bono thinking*. Penguin Books.

Deng, R., Jiang, M., Yu, X., Lu, Y., & Liu, S. (2025). Does ChatGPT enhance student learning? A systematic review and meta-analysis of experimental studies. *Computers & Education, 227*, 105224.

Dimeli, M., & Kostas, A. (2025). The role of Chatgpt in education: Applications and challenges-insights from a systematic review. *Journal of Information Technology Education: Research, 24*, 1–30.

Ennis, H. R. (2015). Critical thinking: A streamlined conception. In M. Davies & R. Barnett (Eds.), *The Palgrave handbook of critical thinking in higher education* (pp. 31–47). Palgrave Macmillan.

Facione, P. (1990). Critical thinking: A statement of expert consensus for purposes of educational assessment and instruction. *(The Delphi Report)*. American Philosophical Association. https://files.eric.ed.gov/fulltext/ED315423.pdf

Fadillah, M. A., Usmeldi, U., & Asrizal, A. (2024). The role of ChatGPT and higher-order thinking skills as predictors of physics inquiry. *Journal of Baltic Science Education, 23*(6), 1178–1192.

Fang, J. W., Guo, X. G., Meng, X. P., Hwang, G. J., & Tu, Y. F. (2025). *Realization of Para social interaction theory in steam education with an AI pedagogical agent: Insights from learning performance, collaboration tendency, and problem-solving tendency*. Advance online publication.

Feuerriegel, S., Hartmann, J., Janiesch, C., & Zschech, P. (2023). *Generative AI*. Advance online publication.

Gerlich, M. (2025). AI tools in society: Impacts on cognitive offloading and the future of critical thinking. *Societies, 15*(1), 6.

Grover, S., & Pea, R. (2013). Computational thinking in K–12: A review of the state of the field. *Educational Researcher, 42*(1), 38–43.

Guo, Q., Zhen, J., Wu, F., He, Y., & Qiao, C. (2025). Can students make STEM progress with the large language models (LLMs)? An empirical study of LLMs integration within middle school science and engineering practice. *Journal of Educational Computing Research, 63*(2), 372–405.

Harari, Y. N. (2018). *21 lições para o século XXI* (2nd ed.). Elsinore.

Haraway, J. D. (2016). *Staying with the trouble: Making kin in the Chthulucene.* Duke University Press.

Hsu, Y.-C., & Ching, Y.-H. (2023). Generative artificial intelligence in education. *Part One: The Dynamic Frontier. TechTrends, 67*(4), 603–607.

Jang, H., & Choi, H. (2025). A double-edged sword: Physics educators' perspectives on utilizing ChatGPT and its future in classrooms. *Journal of Science Education and Technology, 34*(2), 267–283.

Kasneci, E., Sessler, K., Küchemann, S., Bannert, M., Dementieva, D., Fischer, F., et al. (2023). ChatGPT for good? On opportunities and challenges of large language models for education. *Learning and Individual Differences, 103*, 102274.

Kunnath, A. J., & Botes, W. (2025). Transforming science education with artificial intelligence: Enhancing inquiry-based learning and critical thinking in south African science classrooms. *Eurasia Journal of Mathematics, Science and Technology Education, 21*(6).

Lin, Y.-F., Yang, F.-Y., Wu, J.-S., Yeh, C. Y. C., Liao, C.-Y., & Chan, T.-W. (2025). Enhancing students' authentic mathematical problem-solving skills and confidence through error analysis of GPT-4 solutions. *Research and Practice in Technology Enhanced Learning, 20*, 34.

Lipman, M. (1995). *O Pensar na Educação* (A. Paulo, Ed.). Editora Vozes, Ltda.

Liu, J., Sun, D., Sun, J., Wang, J., & Yu, P. L. H. (2025). Designing a generative AI enabled learning environment for mathematics word problem solving in primary schools: Learning performance, attitudes and interaction. *Computers and Education: Artificial Intelligence, 9*, 100438.

Mollick, E. (2024). *Co-intelligence: Living and working with AI.* Hachette Book Group.

Mollick, E., & Mollick, L. (2023). Assigning AI: Seven approaches for students with prompts. Working paper/Guidance note. https://www.researchgate.net/publication/371728434_Assigning_AI_Seven_Approaches_for_Students_with_Prompts

NGSS Lead States. (2013). *Next generation science standards: For states, by states.* The National Academies Press.

OECD. (2021). *AI and the future of skills, volume 1: Capabilities and assessments.* OECD Publishing.

OECD. (2023). *Putting AI to the test: How does the performance of GPT and 15-year-old students in PISA compare?* OECD education spotlights, no. 6. OECD publishing.

Paul, R., & Elder, L. (2018). *The thinker's guide to the nature and functions of critical & creative thinking.* Foundation for Critical Thinking.

Sardi, J., Darmansyah, Candra, O., Yuliana, D. F., Habibullah, Putra Yanto, D. T., & Eliza, F. (2025). How generative AI influences students' self-regulated learning and critical thinking skills? A systematic review. International Journal of Engineering Pedagogy (iJEP), 15(1), 94–108.

Sirisathitkul, C., Sirisathitkul, Y., Wattanasit, K., & Nisoa, M. (2025). Converging objects-to-think-with: Exploring microwave ovens through inquiry-based learning with ChatGPT. *Research in Science & Technological Education.*

Stadler, M., Bannert, M., & Sailer, M. (2024). Cognitive ease at a cost: LLMs reduce mental effort but compromise depth in student scientific inquiry. *Computers in Human Behavior, 160*, 108386.

Tang, K. S., & Putra, G. B. S. (2025). Generative AI as a dialogic partner: Enhancing multiple perspectives, reasoning, and argumentation in science education with customized Chatbots. *Journal of Science Education and Technology.*

Tong, D., Jin, B., Tao, Y., Ren, H., Atiquil Islam, A. Y. M., & Bao, L. (2025). Exploring the role of human-AI collaboration in solving scientific problems. *Physical Review Physics Education Research, 21*(1), 010149.

UNESCO. (2021). *AI and education: Guidance for policy-makers.* UNESCO.

UNESCO. (2023). *Guidance for generative AI in education and research*. UNESCO.

UNESCO. (2024a). *AI competency framework for students*. UNESCO.

UNESCO. (2024b). *AI competency framework for teachers*. UNESCO.

UNESCO. (2025). *AI and the future of education: Disruptions, dilemmas and directions*. UNESCO.

Whittemore, R., & Knafl, K. (2005). The integrative review: Updated methodology. *Journal of Advanced Nursing, 52*(5), 546–553.

Xuan, S. H., Nguyen, A. T., Nguyen, T., Nguyen, L., Nguyen, H., Pham, N., Phung, T., Ngo, B., Nguyen, V., Nguyen, M., Tran, T., Le, T., Nguyen, K., & Fnu, P. (2025). Evaluating the impact of generative AI in mathematics education: A comparative study in Vietnamese high schools. Human Behavior and Emerging Technologies, *Volume 2025*, Article ID 8886206, 23 pages.

Yunianto, W., Galic, S., & Lavicza, Z. (2024). Exploring computational thinking in mathematics education: Integrating ChatGPT with GeoGebra for enhanced learning experiences. *International Journal of Education in Mathematics, Science and Technology, 12*(6), 1451–1470.

Chapter 10
Integrating Generative AI into Science Teaching: Preparing Pre-Service Science Teachers for the AI-Driven STEM Classroom

Tarik El Fathi, Driss Lamri, and El Mehdi Al Ibrahmi

10.1 Introduction

Recent advances in artificial intelligence (AI), driven by breakthroughs in deep learning architectures and the development of large language models (LLMs), have transformed educational practices (Akhtar, 2024; Xing et al., 2025). Generative AI refers to machine-learning systems capable of creating original text, images, or other media in response to natural-language prompts. Models such as ChatGPT, Claude, and DeepSeek exhibit advanced capacities for natural-language interaction, conceptual reasoning, and adaptive dialogue, enabling learners to explore scientific ideas dynamically and conversationally (Al Shloul et al., 2024; Alfredo et al., 2024; Ali et al., 2024). Within K–12 science and teacher-education programs, these tools offer unprecedented opportunities to personalize inquiry, address misconceptions, and support reflective learning (Ammar et al., 2024; Castro et al., 2025; Lee & Kwon, 2024). By simulating Socratic questioning and providing real-time feedback, generative AI enhances inquiry-based learning and fosters critical thinking in STEM classrooms (Gregorcic et al., 2024; Hu & Li, 2023).

However, this technological transformation also introduces significant pedagogical and ethical challenges. While generative AI can generate coherent explanations, it operates through probabilistic inference rather than genuine conceptual understanding, thereby increasing the risk of superficial reasoning and bias (Miao & Holmes, 2023; Fuller & Barnes, 2024; Luo (Jess), 2024). Learners and pre-service

T. El Fathi (✉) · E. M. Al Ibrahmi
Faculty of Science, Ibn Tofail University, Kenitra, Morocco
e-mail: tarik.elfathi@uit.ac.ma

D. Lamri
Regional Center for Education and Training Professions, Rabat-Sale-Kenitra, Morocco

© The Author(s), under exclusive license to Springer Nature Switzerland AG 2026

M. Alkış Küçükaydın (ed.), *Preparing K-12 Pre-Service Science Teachers for the Age of Artificial Intelligence*, Advances in Technology-Rich Science Education 4,
https://doi.org/10.1007/978-3-032-18030-8_10

Table 10.1 Instructional Stages and Cognitive Functions of the CILP Framework

Stage	Description	Teacher Role	Learning Outcome
1. Prompting	Learners formulate initial questions about a scientific concept.	Model and scaffold effective inquiry questions.	Activation of prior knowledge; cognitive engagement.
2. AI-dialogue	Students interact with ChatGPT (or an equivalent LLM) to explore explanations.	Guide questioning, ensure conceptual accuracy, and monitor engagement.	Exposure to multiple perspectives and reasoning approaches.
3. Reflection	Learners evaluate AI responses for accuracy, coherence, and relevance.	Facilitate peer discussion; encourage evidence-based critique.	Metacognitive awareness and self-assessment.
4. Conceptual integration	Students synthesize validated AI insights with experiments, textbooks, or data.	Support mapping of concepts across representations and modalities.	Deep conceptual understanding; coherence across ideas.
5. Evaluation	Students and teachers assess conceptual progress and persistent misconceptions.	Provide formative feedback and foster ethical reflection on AI use.	Consolidated understanding and improved reasoning.

teachers who are unfamiliar with AI limitations may misinterpret its outputs or accept them as authoritative, weakening epistemic agency and reflective judgment (Chan & Lee, 2023; Kim, 2024). These concerns are particularly salient in K–12 teacher education, where future educators must learn to integrate AI as both a cognitive tool and a subject of critical inquiry (Bowe et al., 2025; Kim, 2024; Wu et al., 2024).

The Constructivist Inquiry-Based Learning Prompting (CILP) Framework, introduced in this chapter, directly addresses these tensions. Grounded in constructivist learning theory and inquiry-based science education (Hannafin et al., 1997; Niemi, 2021), the framework positions generative AI as a dialogic partner that complements rather than replaces teacher facilitation. CILP organizes human–AI interaction into iterative cycles of prompting, dialogue, reflection, conceptual integration, and evaluation, guiding pre-service teachers to engage critically with AI outputs, refine prompts, and analyze reasoning processes. By linking prompt engineering with inquiry pedagogy, the framework cultivates epistemic awareness, conceptual understanding, and ethical reflection (El Fathi et al., 2025).

For K–12 pre-service science teachers, AI integration represents more than a technical skill—it constitutes an emerging domain of pedagogical literacy (Tan & Tang, 2025; Wu et al., 2024). Teachers must learn to design AI-supported inquiry tasks, diagnose misconceptions, and facilitate discussions about evidence and bias. Generative AI's immediacy and adaptability make it a powerful scaffold for conceptual understanding; however, its effective use depends on teachers' capacity to balance innovation with critical oversight (El Fathi et al., 2025; Zhai et al., 2024). The CILP

Framework offers a structured pathway for achieving that balance, equipping future educators to employ AI responsibly while fostering inquiry, reflection, and scientific reasoning in K–12 classrooms.

10.2 Chapter Aims and Scope

This chapter introduces and empirically validates the Constructivist Inquiry-Based Learning Prompting (CILP) Framework, a pedagogical model designed to prepare pre-service science teachers to integrate generative artificial intelligence (AI) both ethically and effectively into inquiry-based instruction. Grounded in constructivist learning theory and the principles of inquiry-based science education, the CILP Framework provides a systematic structure for designing and facilitating AI-supported inquiry experiences that ensure responsible and equitable AI use.

It organizes human–AI interaction into five iterative stages—Prompting, AI Dialogue, Reflection, Conceptual Integration, and Evaluation—each transforming generative AI from a passive information provider into an active cognitive and instructional partner. By guiding teachers and learners through these stages, the framework promotes intentional, evidence-informed engagement with AI that upholds academic integrity and fosters inclusive participation in scientific reasoning. Through this process, the CILP Framework cultivates learner agency, metacognitive reflection, and conceptual understanding in science classrooms.

The chapter pursues four interrelated aims that link theory, evidence, and pedagogy:

1. To situate generative AI within constructivist and inquiry-based traditions in science education, clarifying its alignment with learner-centered, knowledge-building pedagogies that advance equity and inclusion.
2. To introduce and explain the CILP Framework as a coherent, adaptable model for AI-mediated inquiry that supports scientific reasoning, conceptual change, and ethical reflection.
3. To present empirical evidence demonstrating that CILP-guided engagement with ChatGPT enhances conceptual understanding and reduces misconceptions in complex thermodynamics topics, with all data collected under institutional ethics approval.
4. To translate these findings into strategies for pre-service teacher education, emphasizing AI literacy, prompt-engineering competence, and hybrid (human–AI) instructional design consistent with fairness, transparency, and responsible AI use.

Structured across eight sections, the chapter first outlines the theoretical foundations of constructivist and inquiry-based learning, then introduces the CILP Framework as a synthesis of these perspectives. Subsequent sections describe the validation study—its participants, data sources, and analytic approach—followed by key findings related to conceptual understanding, metacognition, and ethical reasoning.

The chapter concludes with implications for teacher education and curriculum design, arguing that structured, reflective engagement with generative AI can transform science learning by preparing educators to design and evaluate AI-enhanced environments within the evolving landscape of AI-integrated STEM education.

10.3 Theoretical and Pedagogical Foundations

10.3.1 Constructivist Learning Theory and Inquiry-Based Pedagogy

Constructivist learning theory asserts that knowledge is actively constructed through experience, reflection, and dialogue rather than passively received (Piaget, 1976; Vygotsky, 1978). Inquiry-based pedagogy extends this principle into classroom practice by engaging learners in scientific reasoning processes such as posing questions, collecting evidence, and constructing explanations. Frameworks such as the 5E and 7E instructional models operationalize these inquiry cycles through stages of engagement, exploration, and evaluation (Bybee, 2006, 2016). In K–12 pre-service teacher education, these frameworks provide foundational structures for designing inquiry lessons that integrate conceptual, procedural, and epistemic dimensions of science learning.

Constructivist approaches have also informed the development of innovative pedagogies such as simulation-based inquiry (e.g., PhET simulations) and virtual reality (VR) laboratories, which support reasoning about complex phenomena like thermodynamics and energy transfer (Diab et al., 2024; Santilli et al., 2025). For future science teachers, these methods illustrate how inquiry can be enhanced by technology—whether AI, VR, or simulations—without displacing the essential human facilitation required for conceptual change.

10.3.2 Generative AI in STEM Education: Potentials and Challenges

Generative AI has transformed STEM pedagogy by enabling personalized, adaptive, and interactive learning experiences (Chan & Lee, 2023; Eager & Brunton, 2023). Tools such as ChatGPT provide real-time feedback and context-sensitive explanations, supporting cognitive tasks ranging from hypothesis testing to argumentation. In physics and chemistry education, AI-mediated dialogue has been shown to reduce misconceptions and enhance understanding of complex concepts such as entropy and thermodynamic systems (Demirezen et al., 2023; El Fathi et al., 2025; Elmas et al., 2024). For K–12 pre-service teachers, these findings highlight the potential of AI to scaffold reasoning and promote inquiry-driven engagement when coupled with reflective instruction.

Nevertheless, substantial limitations persist. Generative AI systems may reproduce bias, oversimplify complex ideas, or produce plausible yet inaccurate explanations (Akgun & Greenhow, 2022; OECD, 2023). Overreliance on AI can weaken critical thinking and problem-solving skills—competencies that are central to STEM literacy and teacher expertise (Aslam et al., 2023; Cao et al., 2023; Pinar et al., 2025). Ethical issues surrounding data privacy, intellectual authorship, and equitable access further complicate classroom integration (Akgun & Greenhow, 2022; Nguyen et al., 2023). For teacher education programs, these challenges underscore the need for professional development that prepares educators to evaluate AI critically, design balanced curricula, and model responsible use for students (Fuller & Barnes, 2024; Karataş et al., 2025).

10.3.3 Prompt Engineering and Hybrid Human–AI Inquiry

Prompt engineering—the deliberate design and refinement of natural-language inputs to guide AI responses—has emerged as a central pedagogical competence in AI-supported classrooms (Tassoti, 2024; Wan & Chen, 2024). This practice aligns closely with inquiry-based learning by enabling teachers and students to frame investigable questions, analyze responses, and iteratively refine both prompts and understanding. In STEM education, carefully constructed prompts enhance the precision, coherence, and conceptual depth of AI-generated explanations (Bewersdorff et al., 2023; Wan & Chen, 2024). For K–12 pre-service teachers, prompt engineering functions as both a metacognitive and instructional strategy, training them to guide AI toward productive reasoning while modeling inquiry for their future students.

Hybrid human–AI inquiry positions AI as a collaborative partner in meaning-making rather than a replacement for human instruction (Brusilovsky, 2024; Giordano et al., 2024). Teachers mediate these interactions by validating evidence, contextualizing explanations, and fostering ethical reflection. Learners, in turn, compare AI-generated responses with empirical data, simulations, and authoritative resources, thereby strengthening conceptual integration. In teacher education, this approach cultivates the epistemic and ethical literacies necessary for frameworks such as CILP, where AI functions as a dialogic scaffold for inquiry-based science learning. This orientation aligns with distributed cognition perspectives (Chan & Lee, 2023; Küchemann et al., 2024), positioning AI as part of an expanded cognitive ecosystem rather than an external instructional tool.

10.4 The CILP Framework: Constructivist Inquiry-Based Learning Prompting

10.4.1 Framework Overview and Theoretical Foundations

The Constructivist Inquiry-Based Learning Prompting (CILP) Framework is a recently developed pedagogical model that guides the ethical and effective integration of generative artificial intelligence (AI) tools into STEM education (El Fathi et al., 2025). Developed by Tarik El Fathi, the framework is grounded in the core principles of constructivist learning theory (Hannafin et al., 1997; Lebow, 1993). Drawing on these foundations, CILP conceptualizes learning as an active, iterative, and socially mediated process of knowledge construction, reflection, and meaning-making. It emphasizes that understanding develops through exploration, dialogue, and evaluation rather than through passive information transmission. Accordingly, learners construct meaning by generating, testing, and refining ideas through interactions with both human and digital partners.

At its core, the CILP Framework integrates inquiry-based learning with prompt engineering—the systematic design, testing, and refinement of natural-language prompts to engage generative AI models such as ChatGPT in scientifically oriented reasoning. However, CILP extends beyond procedural prompt design by embedding metacognitive monitoring, self-regulated learning, and epistemic agency within the AI-mediated inquiry process. This structure encourages learners to analyze, critique, and refine both their prompts and the resulting AI outputs. Through this approach, the framework repositions AI as a cognitive collaborator in meaning-making rather than a mere information provider, countering tool-centric uses of AI that foster superficial engagement or uncritical acceptance of AI-generated content.

The CILP Framework also promotes responsible, transparent, and equitable AI use in science education. It encourages teachers and learners to engage critically with AI outputs, ensuring that intellectual integrity, inclusivity, and data ethics are maintained throughout inquiry processes. The framework bridges established constructivist and inquiry-based traditions with emerging domains of AI literacy, data literacy, and critical digital pedagogy. It provides a model for cultivating learners' scientific reasoning, metacognitive regulation, and ethical discernment in AI-enhanced science education.

This chapter empirically examines the framework within a quasi-experimental classroom context. All instructional activities adhered to institutional ethical standards for educational research, including informed consent, data anonymization, and participant confidentiality. Because a formal validation study by the original developer remains in progress, findings are interpreted cautiously and should be considered preliminary rather than definitive.

10.4.2 Instructional Stages and Learning Functions

The CILP Framework operates through a five-stage learning cycle that integrates human and AI dialogue within inquiry-based environments. Each stage corresponds to specific instructional roles and targeted cognitive outcomes, forming a recursive, inquiry-driven process that parallels authentic scientific investigation. The cycle's progression—from question generation to reflection and evaluation—provides a structured scaffold that links inquiry, reasoning, and conceptual integration.

This cyclic process mirrors the scientific method—hypothesis formulation, testing, reflection, and validation—within an AI-mediated inquiry context. Ethical considerations are embedded throughout, guiding learners to attribute sources properly, recognize AI limitations, and engage respectfully in both human–AI and peer dialogue. By aligning inquiry stages with structured AI interaction, CILP transforms generative dialogue into a scaffold for conceptual change, epistemic reflection, and evidence-based reasoning. It fosters dynamic knowledge construction, self-regulation, and ethical awareness, enabling teachers to facilitate AI-enhanced inquiry that remains learner-centered, inclusive, and methodologically rigorous.

10.5 Empirical Validation of the CILP Framework

10.5.1 Study Context and Participants

This quasi-experimental investigation was conducted during the autumn semester of the 2024–2025 academic year and involved 120 first-year students enrolled at the National School of Applied Sciences, Ibn Tofail University (Kenitra, Morocco). Participants were recruited through convenience sampling and were enrolled in an introductory thermodynamics course, a core component of the preparatory-cycle curriculum. The cohort comprised 54 men and 66 women aged 19–21 years and was divided into two groups matched for prior academic achievement and demographic characteristics.

The experimental group ($n = 60$) engaged with the freely accessible version of ChatGPT integrated within the CILP Framework, whereas the control group ($n = 60$) received traditional instruction without AI support. To ensure instructional equivalence and curricular alignment, both groups participated in the same lectures, laboratory sessions, and assessment schedules throughout the intervention period. The intervention spanned eight weeks, consisting of weekly two-hour tutorial sessions. Experimental sessions were conducted in a computer laboratory equipped with internet-enabled workstations and emphasized prompt construction, dialogic engagement with AI, and reflective analysis aligned with the CILP inquiry cycle. The control sessions took place in a conventional classroom environment, focusing on problem-solving, guided exercises, and instructor-led explanations following established pedagogical practices.

Instructional time, learning objectives, and assessment procedures were held constant across groups, with AI integration serving as the sole experimental variable. The intervention addressed the following research question:

To what extent does integrating ChatGPT as a pedagogical scaffold within the CILP Framework enhance first-year students' conceptual understanding of thermodynamics and facilitate the identification and reduction of misconceptions?

10.6 Findings

The findings provide convergent evidence of conceptual improvement and a systematic reduction in misconceptions under the CILP condition. Students in the experimental group achieved significantly higher posttest scores ($M = 22.53$) than those in the control group ($M = 18.60$). On average, CILP participants improved by 10.31 points, while control participants improved by 6.25 points. A Mann–Whitney U test confirmed that this difference was statistically significant ($p < .001$), with a moderate-to-large effect size (Cohen's $d = 0.61$). Within-group analyses further revealed more pronounced pretest-to-posttest gains for the experimental condition, corroborating the positive impact of AI-mediated inquiry.

To examine the coherence of conceptual understanding, a principal component analysis (PCA) was performed. Whereas the control group's conceptual structure remained fragmented across thermodynamic domains (e.g., ideal gas, entropy, and cyclic processes), the experimental group exhibited a single, integrated factor—"General Thermodynamics Understanding"—accounting for 71.4% of the variance. This outcome suggests that iterative AI feedback fostered conceptual synthesis across themes, particularly regarding relationships among work, heat, and internal energy.

Regarding misconception reduction, chi-square analyses revealed significant improvement in the experimental group across key conceptual barriers, particularly in understanding entropy and irreversible processes ($p < .001$). Conversely, the control group demonstrated only marginal gains, with persistent misconceptions concerning the ideal gas law and cyclic transformations. Cluster analyses further indicated that students entering with high misconception levels benefited from AI-supported dialogue but required more targeted prompting to achieve comparable progress. Students with moderate or low misconception levels exhibited substantial improvement supported by iterative feedback and reflection. However, quantitative reasoning tasks—such as those involving engine efficiency and reversible processes—remained challenging, underscoring limitations in ChatGPT's numerical problem-solving capabilities.

In summary, the empirical findings validate the CILP Framework as both a pedagogical innovation and a theoretical contribution to inquiry-based science education. The framework operationalizes constructivist principles by integrating dialogic engagement, iterative reflection, and AI-mediated scaffolding into instructional design. Collectively, these processes promote measurable conceptual transformation in complex scientific domains. The results further demonstrate that structured human–AI collaboration enhances conceptual understanding and reduces misconceptions without diminishing teacher expertise. Instead, AI functions as a dialogic and cognitive partner within a carefully scaffolded inquiry process, reinforcing the co-construction of knowledge central to the CILP model.

10.7 From Research to Practice

Building on the empirical validation of the Constructivist Inquiry-Based Learning Prompting (CILP) Framework, this section translates the research findings into pedagogical insights and practical approaches for K–12 pre-service teacher education and science classroom implementation. It examines how CILP informs AI literacy, teacher preparation, and the design of inquiry-driven learning environments that integrate generative AI responsibly and effectively.

10.8 Implications for Teacher Education and AI Literacy

The findings underscore the importance of pre-service and in-service teacher education programs that prioritize AI literacy, prompt engineering, and hybrid instructional design. Teachers must develop a conceptual understanding of generative AI—its capacities for conceptual explanation and its limitations in numerical or procedural reasoning. The CILP Framework functions as a pedagogical scaffold that enables educators to design AI-mediated lessons promoting inquiry, reflection, and critical thinking. Training programs should include structured activities in which teachers craft open-ended prompts, evaluate AI-generated responses for accuracy, coherence, and bias, and guide students in refining their inquiries. Equally essential are reflective discussions on ethics, data privacy, intellectual authorship, and the risks of overreliance on AI-generated content.

Teacher education programs should also equip K–12 science teachers to integrate AI tools alongside complementary pedagogies such as laboratory experimentation, computer simulations, and collaborative learning. Generative AI can, for instance, support the exploration of molecular structures or reaction mechanisms, while teacher-led discussions ensure conceptual coherence and epistemic rigor. Ultimately, educators must design hybrid environments that leverage AI's dynamic feedback while preserving the irreplaceable value of human mentorship and pedagogical judgment.

10.9 Preparing AI-Ready Pre-Service Science Teachers

Preparing pre-service science teachers (PSTs) for AI-integrated classrooms requires the development of competencies that extend beyond traditional content knowledge. Drawing on the CILP study and current literature, four interrelated areas of competence emerge: (1) prompt engineering for pedagogical purposes, (2) AI literacy and interpretive reasoning, (3) hybrid instructional design, and (4) digital ethics and equity awareness. PSTs must learn to design and scaffold prompts that elicit accurate, conceptually rich explanations, anticipate likely AI responses, and critically assess their validity. They also require a foundational understanding of how generative models function—including data constraints, probabilistic reasoning, and bias detection—while guiding students to verify AI-generated content.

In hybrid learning contexts, teachers must determine when and how to introduce AI so that students engage with it as a cognitive partner rather than an unquestioned authority. Ethical awareness is equally vital: PSTs should address issues of privacy, authorship, and equitable access while recognizing how sociocultural and technological factors shape AI use in classrooms. To cultivate these skills, science teacher education programs should embed CILP-aligned modules within methods courses, allowing PSTs to design inquiry prompts, interact with AI models, analyze outputs, and reflect on pedagogical implications. Additional strategies—such as AI-integrated microteaching, reflective portfolios, and case-based analyses—can further strengthen metacognitive and ethical development.

Accessibility challenges can be mitigated by providing low-barrier AI platforms, offline exemplars, and collaborative peer activities, particularly in under-resourced contexts. Through these efforts, pre-service teachers gain critical data literacy and retain instructional agency in AI-rich environments, positioning themselves as reflective practitioners who integrate generative AI responsibly, equitably, and constructively.

10.10 Strategies for Integrating Generative AI in Science Classrooms

Building on these foundations, effective K–12 classroom integration requires explicit, scaffolded instructional strategies. Teachers should begin with low-barrier entry points—using freely accessible AI tools for concept checks or exploratory questioning—and progressively increase task complexity to prevent cognitive overload. The CILP Framework encourages students to craft inquiry-driven prompts aligned with learning objectives, evaluate AI responses against empirical evidence, and refine their reasoning through reflection and peer dialogue.

Integrating AI into formative assessment allows educators to generate practice problems, quizzes, or Socratic dialogues that address common misconceptions, provided the outputs are carefully reviewed for accuracy and bias. Combining AI with laboratory work, simulations, and collaborative problem-solving helps connect virtual inquiry to empirical investigation. For instance, after exploring a thermodynamics concept using ChatGPT, students may conduct experiments or analyze PhET simulations to verify and discuss the results. Teachers should monitor AI use closely, offering formative feedback and targeted scaffolds when misconceptions persist.

Cultivating AI literacy among students is equally critical. Learners should understand how generative models operate—their mechanisms, biases, and limitations—to develop critical digital literacy and ethical awareness. Guided use of AI enables students to engage with it as a reasoning partner in inquiry-based problem-solving rather than as an unquestioned authority. Collectively, these pedagogical strategies and preparation models illustrate how CILP unites AI literacy, inquiry-based pedagogy, and conceptual understanding in STEM education. Realizing this potential requires sustained empirical research, iterative design, and critical reflection on ethics and inclusivity.

Taken together, these insights position the CILP Framework as a bridge between research and practice, equipping K–12 pre-service science teachers to implement AI-enhanced inquiry pedagogy responsibly and effectively. These outcomes lay the groundwork for the next section, which explores future directions, research opportunities, and pathways for scaling and replicating the CILP Framework across science disciplines.

10.11 Future Directions and Replicability

Although this chapter focused empirically on thermodynamics instruction within engineering education, the Constructivist Inquiry-Based Learning Prompting (CILP) Framework is inherently domain-transcendent and adaptable across K–12 and higher education science contexts. Its central principle—structured inquiry through iterative prompting and dialogic reflection—extends naturally to disciplines such as biology, chemistry, physics, and environmental science, where generative AI can support conceptual modeling, analogy building, and reflective reasoning about complex systems. The framework's modular design accommodates diverse curricula, pedagogies, and resource levels, enabling educators to scale it from short interventions to comprehensive instructional units aligned with inquiry-based science education, problem-based learning, and national science standards.

Sustaining and expanding this work will require cross-institutional collaboration among K–12 schools and teacher-education programs to co-develop prompt repositories, share lesson exemplars, and conduct multi-site design-based research grounded in open-science practices such as preregistration and data sharing. These collaborations—aligned with guidelines from organizations such as UNESCO and the OECD on responsible AI—can enhance transparency and replicability while strengthening capacity in under-resourced settings (Miao & Holmes, 2023; OECD, 2023). However, model drift, or updates to large language models between 2024 and 2025, may influence reproducibility; therefore, future studies should explicitly document the model versions employed and contextualize results accordingly.

To ensure methodological rigor and replicability, future research should adopt design-based methodologies combined with mixed-method evaluations that capture shifts in epistemological beliefs, AI-literacy development, and pedagogical adaptation across diverse educational contexts. Such studies can illuminate how teachers' confidence, agency, and epistemic framing of AI evolve over time, as well as how generative tools can be integrated ethically and equitably across classroom environments. Researchers should also investigate longitudinal outcomes, examining how iterative prompt refinement and reflective AI dialogue influence conceptual retention, reasoning strategies, and the transfer of inquiry practices across STEM domains.

Ultimately, the framework's promise lies not in technological novelty but in educators' capacity to use generative AI as a partner in cultivating curiosity, conceptual understanding, and ethical scientific literacy. Grounded in sound pedagogy, the CILP Framework positions generative AI as a catalyst for innovation—rather than disruption—in preparing future-ready, inquiry-driven K–12 science teachers and learners.

10.12 Conclusion

Building upon the practical implications outlined in the preceding section, the integration of generative artificial intelligence (AI) into science education is no longer speculative—it has become a pedagogical imperative. As demonstrated throughout this chapter, when anchored in constructivist and inquiry-based paradigms, generative tools such as ChatGPT can enhance conceptual understanding, address persistent misconceptions, and foster reflective engagement with complex scientific ideas. The Constructivist Inquiry-Based Learning Prompting (CILP) Framework introduced in this work provides a replicable and adaptable model for integrating AI into science instruction. Grounded in constructivist learning theory, CILP reconceptualizes generative AI as a cognitive partner in inquiry, equipping K–12 pre-service science teachers to design, refine, and reflect upon AI interactions that promote epistemic agency, conceptual coherence, and pedagogical growth.

Empirical validation of the framework revealed significant learning gains in thermodynamics and identified interaction patterns distinguishing reflective, judicious AI use from uncritical or excessive prompting. These findings underscore the need for explicit preparation in prompt engineering, AI and data literacy, and hybrid (human–AI) instructional design as essential competencies in science teacher education. Through structured human–AI collaboration, learners develop deeper understanding and enhanced reasoning capacities while teachers retain their central role as facilitators of inquiry and ethical reflection.

Equally vital are the ethical, cultural, and equity-based principles that guide responsible AI integration. From low-barrier implementation in resource-constrained settings to the critical evaluation of AI-generated knowledge, effective practice demands both technical proficiency and ethical discernment. The continued evolution of frameworks such as CILP will depend on cross-institutional collaboration, longitudinal research, and sustained dialogue among educators, researchers, and policymakers. Such efforts ensure that innovation in AI integration remains anchored in pedagogical integrity, inclusivity, and care rather than technological determinism.

When guided by sound pedagogy and informed by empirical evidence, generative AI becomes not a substitute for teachers but a powerful ally in cultivating conceptual understanding, critical reasoning, and scientific literacy. In preparing K–12 pre-service science teachers for AI-driven STEM classrooms, the goal extends beyond equipping them with new tools—it is about empowering them to shape the future of teaching and learning itself. The CILP Framework thus represents both a practical pathway and a theoretical commitment to integrating generative AI in ways that sustain inquiry, promote ethical reflection, and advance equitable access to high-quality science education.

References

Akgun, S., & Greenhow, C. (2022). Artificial intelligence in education: Addressing ethical challenges in K-12 settings. *AI and Ethics, 2*(3), 431–440. https://doi.org/10.1007/s43681-021-00096-7

Akhtar, Z. B. (2024). Unveiling the evolution of generative AI (GAI): A comprehensive and investigative analysis toward LLM models (2021–2024) and beyond. *Journal of Electrical Systems and Information Technology, 11*(1), 22. https://doi.org/10.1186/s43067-024-00145-1

Al Shloul, T., Mazhar, T., Abbas, Q., Iqbal, M., Ghadi, Y. Y., Shahzad, T., Mallek, F., & Hamam, H. (2024). Role of activity-based learning and ChatGPT on students' performance in education. *Computers and Education: Artificial Intelligence, 6*, 100219. https://doi.org/10.1016/j.caeai.2024.100219

Alfredo, R., Echeverria, V., Jin, Y., Yan, L., Swiecki, Z., Gašević, D., & Martinez-Maldonado, R. (2024). Human-centred learning analytics and AI in education: A systematic literature review. *Computers and Education: Artificial Intelligence, 6*, 100215. https://doi.org/10.1016/j.caeai.2024.100215

Ali, D., Fatemi, Y., Boskabadi, E., Nikfar, M., Ugwuoke, J., & Ali, H. (2024). ChatGPT in teaching and learning: A systematic review. *Education Sciences, 14*(6), 643. https://doi.org/10.3390/educsci14060643

Ammar, M., Al-Thani, N. J., & Ahmad, Z. (2024). Role of pedagogical approaches in fostering innovation among K-12 students in STEM education. *Social Sciences & Humanities Open, 9*, 100839. https://doi.org/10.1016/j.ssaho.2024.100839

Aslam, S., Alghamdi, A. A., Abid, N., & Kumar, T. (2023). Challenges in implementing STEM education: Insights from novice STEM teachers in developing countries. *Sustainability, 15*(19), 14455. https://doi.org/10.3390/su151914455

Bewersdorff, A., Seßler, K., Baur, A., Kasneci, E., & Nerdel, C. (2023). Assessing student errors in experimentation using artificial intelligence and large language models: A comparative study with human raters. *Computers and Education: Artificial Intelligence, 5*, 100177. https://doi.org/10.1016/j.caeai.2023.100177

Bowe, K. A., Altindis, N., Bauer, C. F., & Aikens, M. L. (2025). Cognitive resources drawn from different STEM disciplines for interpreting a titration experiment logistic graph: A case study of student thinking. *Disciplinary and Interdisciplinary Science Education Research, 7*(1), 20. https://doi.org/10.1186/s43031-025-00140-0

Brusilovsky, P. (2024). AI in education, learner control, and human-AI collaboration. *International Journal of Artificial Intelligence in Education, 34*(1), 122–135. https://doi.org/10.1007/s40593-023-00356-z

Bybee, R. W. (2006). Scientific inquiry and science teaching. In L. B. Flick & N. G. Lederman (Eds.), *Scientific inquiry and nature of science: Implications for teaching, learning, and teacher education* (pp. 1–14). Springer Netherlands. doi:https://doi.org/10.1007/978-1-4020-5814-1_1.

Bybee, R. W. (2016). *The BSCS 5E instructional model: Creating teachable moments.* SAGE Publications.

Cao, C., Ding, Z., Lee, G.-G., Jiao, J., Lin, J., & Zhai, X. (2023). *Elucidating STEM concepts through generative.* A Multi-modal Exploration of Analogical Reasoning. https://doi.org/10.48550/arXiv.2308.10454

Castro, S. K., Brown, J. C., & Crippen, K. J. (2025). Fostering critical consciousness: A systematic review of K-12 teachers' integrations of sociopolitical issues in science and mathematics classrooms. *Disciplinary and Interdisciplinary Science Education Research, 7*(1), 7. https://doi.org/10.1186/s43031-025-00126-y

Chan, C. K. Y., & Lee, K. K. W. (2023). The AI generation gap: Are gen Z students more interested in adopting generative AI such as ChatGPT in teaching and learning than their gen X and millennial generation teachers? Smart Learning Environments, 10(1), 60. doi:https://doi.org/10.1186/s40561-023-00269-3.

Demirezen, M. U., Yilmaz, O., & Ince, E. (2023). New models developed for detection of misconceptions in physics with artificial intelligence. *Neural Computing and Applications, 35*(12), 9225–9251. https://doi.org/10.1007/s00521-023-08414-2

Diab, H., Daher, W., Rayan, B., Issa, N., & Rayan, A. (2024). Transforming science education in elementary schools: The power of PhET simulations in enhancing student learning. *Multimodal Technologies and Interaction, 8*(11), 105. https://doi.org/10.3390/mti8110105

Eager, B., & Brunton, R. (2023). Prompting higher education towards AI-augmented teaching and learning practice. *Journal of University Teaching & Learning Practice, 20*(5). https://doi.org/10.53761/1.20.5.02

El Fathi, T., Saad, A., Larhzil, H., Lamri, D., & Al Ibrahmi, E. M. (2025). Integrating generative AI into STEM education: Enhancing conceptual understanding, addressing misconceptions, and assessing student acceptance. *Disciplinary and Interdisciplinary Science Education Research, 7*(1), 6. https://doi.org/10.1186/s43031-025-00125-z

Elmas, R., Adiguzel-Ulutas, M., & Yılmaz, M. (2024). Examining ChatGPT's validity as a source for scientific inquiry and its misconceptions regarding cell energy metabolism. *Education and Information Technologies.* https://doi.org/10.1007/s10639-024-12749-1

Fuller, M., & Barnes, N. (2024). The impact of ChatGPT on teaching and learning in higher education: Exploring the dual perspectives of participants who were students and teachers. *New Directions for Higher Education, 2024*(207), 31–46. https://doi.org/10.1002/he.20507

Giordano, V., Spada, I., Chiarello, F., & Fantoni, G. (2024). The impact of ChatGPT on human skills: A quantitative study on twitter data. *Technological Forecasting and Social Change, 203*, 123389. https://doi.org/10.1016/j.techfore.2024.123389

Gregorcic, B., Polverini, G., & Sarlah, A. (2024). ChatGPT as a tool for honing teachers' Socratic dialogue skills. *Physics Education, 59*(4), 045005. https://doi.org/10.1088/1361-6552/ad3d21

Hannafin, M. J., Hannafin, K. M., Land, S. M., & Oliver, K. (1997). Grounded practice and the design of constructivist learning environments. *Educational Technology Research and Development, 45*(3), 101–117. https://doi.org/10.1007/BF02299733

Hu, W., & Li, X. (2023). Creativity in STEM education. In R. J. Tierney, F. Rizvi, & K. Ercikan (Eds.), International Encyclopedia of education *(Fourth Edition)* (pp. 64–75). Elsevier. https://www.sciencedirect.com/science/article/pii/B978012818630513009X

Karataş, F., Eriçok, B., & Tanrikulu, L. (2025). Reshaping curriculum adaptation in the age of artificial intelligence: Mapping teachers' AI-driven curriculum adaptation patterns. *British Educational Research Journal, 51*(1), 154–180. https://doi.org/10.1002/berj.4068

Kim, J. (2024). Types of teacher-AI collaboration in K-12 classroom instruction: Chinese teachers' perspective. *Education and Information Technologies.* https://doi.org/10.1007/s10639-024-12523-3

Küchemann, S., Steinert, S., Kuhn, J., Avila, K., & Ruzika, S. (2024). Large language models—Valuable tools that require a sensitive integration into teaching and learning physics. *The Physics Teacher, 62*(5), 400–402. https://doi.org/10.1119/5.0212374

Lebow, D. (1993). Constructivist values for instructional systems design: Five principles toward a new mindset. *Educational Technology Research and Development, 41*(3), 4–16. https://doi.org/10.1007/BF02297354

Lee, S. J., & Kwon, K. (2024). A systematic review of AI education in K-12 classrooms from 2018 to 2023: Topics, strategies, and learning outcomes. *Computers and Education: Artificial Intelligence, 6*, 100211. https://doi.org/10.1016/j.caeai.2024.100211

Luo (Jess), J. (2024). How does GenAI affect trust in teacher-student relationships? Insights from students' assessment experiences. *Teaching in Higher Education, 1–16*. https://doi.org/10.1080/13562517.2024.2341005

Miao, F., & Holmes, W. (2023). Guidance for generative AI in education and research.

Nguyen, A., Ngo, H. N., Hong, Y., Dang, B., & Nguyen, B.-P. T. (2023). Ethical principles for artificial intelligence in education. *Education and Information Technologies, 28*(4), 4221–4241. https://doi.org/10.1007/s10639-022-11316-w

Niemi, H. (2021). AI in learning: Preparing grounds for future learning. *Journal of Pacific Rim Psychology, 15*, 18344909211038105. https://doi.org/10.1177/18344909211038105

OECD. (2023). Common guideposts to promote interoperability in AI risk management. OECD. https://www.oecd-ilibrary.org/science-and-technology/common-guideposts-to-promote-interoperability-in-ai-risk-management_ba602d18-en

Piaget, J. (1976). Piaget's theory. In B. Inhelder, H. H. Chipman, & C. Zwingmann (Eds.), *Piaget and his school: A reader in Developmental Psychology* (pp. 11–23). Springer. https://doi.org/10.1007/978-3-642-46323-5_2

Pinar, F. I. L., Panergayo, A. A. E., Sagcal, R. R., Acut, D. P., Roleda, L. S., & Prudente, M. S. (2025). Fostering scientific creativity in science education through scientific problem-solving approaches and STEM contexts: A meta-analysis. Disciplinary and Interdisciplinary Science Education Research, 7(1), 18. doi:https://doi.org/10.1186/s43031-025-00137-9.

Santilli, T., Ceccacci, S., Mengoni, M., & Giaconi, C. (2025). Virtual vs. traditional learning in higher education: A systematic review of comparative studies. *Computers & Education, 227*, 105214. https://doi.org/10.1016/j.compedu.2024.105214

Tan, Q., & Tang, X. (2025). Unveiling AI literacy in K-12 education: A systematic literature review of empirical research. *Interactive Learning Environments, 1–17*. https://doi.org/10.1080/10494820.2025.2482586

Tassoti, S. (2024). Assessment of students use of generative artificial intelligence: Prompting strategies and prompt engineering in chemistry education. *Journal of Chemical Education, 101*(6), 2475–2482. https://doi.org/10.1021/acs.jchemed.4c00212

Vygotsky, L. S. (1978). Mind in society: Development of higher psychological processes. *Harvard University Press*. https://doi.org/10.2307/j.ctvjf9vz4

Wan, T., & Chen, Z. (2024). Exploring generative AI assisted feedback writing for students' written responses to a physics conceptual question with prompt engineering and few-shot learning. *Physical Review Physics Education Research, 20*(1), 010152. https://doi.org/10.1103/PhysRevPhysEducRes.20.010152

Wu, D., Chen, M., Chen, X., & Liu, X. (2024). Analyzing K-12 AI education: A large language model study of classroom instruction on learning theories, pedagogy, tools, and AI literacy. *Computers and Education: Artificial Intelligence, 7*, 100295. https://doi.org/10.1016/j.caeai.2024.100295

Xing, W., Nixon, N., Crossley, S., Denny, P., Lan, A., Stamper, J., & Yu, Z. (2025). The use of large language models in education. *International Journal of Artificial Intelligence in Education, 35*(2), 439–443. https://doi.org/10.1007/s40593-025-00457-x

Zhai, C., Wibowo, S., & Li, L. D. (2024). The effects of over-reliance on AI dialogue systems on students' cognitive abilities: A systematic review. *Smart Learning Environments, 11*(1), 28. https://doi.org/10.1186/s40561-024-00316-7

Chapter 11
The Long-Term Possibilities and Challenges of Artificial Intelligence in K-12 Science Education

Melahat Çelik Demir and Mustafa Doğru

11.1 Introduction

AI is a transformative technology that enables machines to perform tasks and processes requiring human intelligence. Here, AI first enables machines to perceive their environment, reason, learn from experience, and make autonomous decisions in order to perform the relevant tasks. In addition, AI promotes innovation, increases efficiency, and is seen as a problem-solving mechanism for society thanks to its ability to process large amounts of data (Hua Liu, 2024). This shows that AI has the potential to be used in many areas of society.

Today, with technological advances, many areas are undergoing change and transformation. With advances in technology, it has become inevitable that AI transform many areas (Hutson et al., 2022). One of these areas is education. Indeed, the twenty-first-century understanding of education is being reshaped by technological development and transformation. In this regard, the most critical and current form of transformation is seen as the integration of AI technology into education (Kamalov et al., 2023). In parallel, AI has been seen as the driving force behind change, focusing on students' needs and desires (Tapalova et al., 2022). The rapid advancement of AI technology has mobilized the education community in response to this driving force. In this regard, this situation has been addressed in detail for educators and education assessment experts. For them, AI technologies have become indispensable for improving teaching and learning experiences (Al Darayseh, 2023; Owan et al., 2023).

AI in education is an interdisciplinary field due to its ability to transform teaching and learning environments. The fact that this is particularly applied

M. Ç. Demir (✉) · M. Doğru
Akdeniz University, Antalya, Turkey

M. Doğru
e-mail: mustafadogru@akdeniz.edu.tr

© The Author(s), under exclusive license to Springer Nature Switzerland AG 2026
M. Alkış Küçükaydın (ed.), *Preparing K-12 Pre-Service Science Teachers for the Age of Artificial Intelligence*, Advances in Technology-Rich Science Education 4,
https://doi.org/10.1007/978-3-032-18030-8_11

in science education has increased interest in and popularity of artificial intelligence (Chiu et al., 2023). These results have been observed especially in K-12 science education. Indeed, since science education aims to develop students' scientific thinking, discovery, and problem-solving skills, AI technologies facilitate this process (Almasri, 2024). Therefore, the importance of AI in K-12 education has increased over time, making it crucial for teachers to possess the necessary competencies to effectively and efficiently educate students about AI (Wang et al., 2023).

Considering the importance of AI in K-12 science education, it has been observed that it affects both teachers and students. For example, AI-based learning systems analyze collected data, identify areas for improvement, and create personalized learning opportunities by adjusting the pace and content of lessons according to students' individual performance (Almasri, 2024; Demartini et al., 2024). In addition, the potential of smart learning systems to create customized content, provide students with feedback, and support teachers through automated grading cannot be overlooked (du Boulay, 2019). Studies have found that AI-supported instruction has a positive effect on students' academic achievement (Zhao et al., 2019) and contributes significantly to student performance and learning in science education (Deveci Topal et al., 2021). In addition, a study examined how elementary school teachers used AI in science lessons. The study concluded that teachers need support in using AI and integrating it into education. Another finding was that science classes were the most popular for AI use (AlKanaan, 2022; Shin & Shin, 2020).

AI applications in education enable individuals to learn independently rather than simply acquire, process, and store information. This supports science teachers in considering students' individual differences. In this way, AI applications used by K-12 science teachers improve the quality of education and increase learning efficiency (Al Darayseh, 2023). Therefore, it is necessary to identify the difficulties teachers face when using AI in science lessons in K-12 education. Indeed, resolving the difficulties teachers experience in this process will enable the education and teaching process to progress more rapidly.

Given AI's active role and potential in K-12 education, while positive outcomes may occur, some negative situations may also arise. Indeed, the widespread use of AI in science education does not necessarily mean that teachers will always use it effectively in their lessons (Al Darayseh, 2023). This is because factors such as teachers' interest, skills, readiness, and opportunities also shape the process. Therefore, the use of AI by science teachers in K-12 education does not guarantee effective teaching (United Nations Educational, Scientific, and Cultural Organization, 2019). In addition, it has been observed that, alongside the integration of AI into the general education process, there is a need for extensive research on the sustainability of teachers' AI use, particularly in science education (Al Darayseh, 2023). Therefore, understanding the long-term possibilities and challenges of AI in K-12 science education will provide insight for future research.

11.2 The Long-Term Potential of Artificial Intelligence in K-12 Science Education

11.2.1 Personalized Learning Environments

It is clear that AI technologies have brought about change and transformation in education. These changes and transformations have also affected teaching processes (Luckin et al., 2016). It has been determined that the use of AI in educational processes has many long-term advantages. One of these is that it enables personalized learning environments (Hutson et al., 2022; Ngo Ndjama, 2025; Owan et al., 2023).

Digital learning platforms can personalize and enrich science lessons based on students' learning needs, readiness, academic performance, learning speed, and interest levels (Zhai et al., 2021). They can include activities that will make the student enjoy the lesson while learning. In addition, teachers can shape student assignments according to the students' skills, abilities, and needs and plan the level of difficulty. This personalized learning provided in this way is one of the most important contributions AI makes to the field of science education. Personalized learning environments provided to students increase their academic achievement and raise their level. In the long term, AI-based applications identify students' weaknesses through personalized learning and support their overall development by offering appropriate content and methods (AlKanaan, 2022; Karsenti, 2019). Furthermore, the use of AI in science education makes it more appealing to students with diverse learning styles and needs, as it enables the personalization and development of interactive, engaging learning environments (Almasri, 2024).

11.2.2 Monitoring Student Development and Feedback

One of the biggest challenges teachers face in the education process is systematically monitoring students' academic progress in class. Identifying students with weak academic development and taking appropriate measures for them is considered a necessity. At this point, advanced AI-supported prediction models have become a necessity (Khan et al., 2021). Many advanced prediction models track students' academic performance and can help form conclusions about it (Yang et al., 2017). For example, it has been observed that prediction models can identify students with weak academic performance at the beginning of the school year who are highly likely to fail the course (Asogwa & Oladugba, 2015).

AI technologies enable science educators to track and monitor student progress more effectively. Therefore, this helps science educators to intervene in a targeted manner and provide support to students when necessary (Almasri, 2024). In addition, AI-based teaching systems generate digital progress charts to track students' development and visualize this development process. This allows teachers to keep track of students' strengths and weaknesses on an ongoing basis (Almasri, 2024; Mavroudi et al., 2018; D'Mello & Graesser, 2015). This enables teachers to provide faster feedback on areas where students are lacking and helps students address their shortcomings. This increases students' academic achievement and motivation (Romero & Ventura, 2020).

11.2.3 Data-Driven Teaching and Evaluation

It is clear that AI technologies have brought about change and transformation in education. These changes and transformations have affected not only teaching methods and techniques, but also assessment and evaluation tools and teaching processes (Luckin et al., 2016). It has also been determined that AI-supported educational assessment tools offer numerous advantages. Ensuring the evaluation of the education provided has advantages such as increasing the effectiveness of education, providing individualized feedback to students, increasing access and retention, and reducing education costs (Hutson et al., 2022; Ngo Ndjama, 2025; Owan et al., 2023). Therefore, AI applications are important in evaluating the education provided to students. At this point, traditional measurement and evaluation methods take a back seat. Indeed, traditional measurement and evaluation methods generally do not provide an in-depth examination of students' knowledge levels. This means they measure students' knowledge levels in a limited way, preventing a holistic approach to students' learning. In addition, AI-supported measurement and evaluation systems continuously and in detail reveal students' learning processes and behaviors. This provides students with personalized feedback (Baker & Inventado, 2014). Considering this, the use of AI has become increasingly important over time.

AI is increasingly used in schools for teaching and assessing students. This has led to changes in how students learn. It is evident that AI is a powerful tool for exam assessment and monitoring, as well as for course management. Based on this, integrating AI into educational measurement and evaluation processes can enable more comprehensive and dynamic tracking of student development. In addition, it has been observed that teachers use AI technologies to assess students. This situation provides the teacher with a large dataset, in addition to students' interactions with AI technologies. With this dataset, the teacher can perform learning analytics and evaluate each situation. For example, it allows for the reorganization of teaching methods and techniques, in-class materials, and measurement and evaluation processes. In the long term, this approach may lead to traditional methods being replaced by continuously and dynamically advancing AI-based systems (Ngo Ndjama, 2025).

11.2.4 Experiential Science Education

When considering the profound and revolutionary developments AI is bringing to education, the future of science education is both exciting and thought-provoking, as it will pose challenges. Indeed, it is well known that AI has a transformative effect on both learning and teaching science (Almasri, 2024). The importance of experiments in science cannot be denied. It is not always possible to provide physical experimental environments. At this point, today's groundbreaking AI technologies have produced solutions to this situation.

When examining the effects of AI in science education, the first prominent feature is its ability to simulate scientific experiments. Additionally, it provides students with the opportunity to conduct experiments through virtual laboratories in science classes. This not only allows students to apply and develop their scientific skills but also saves on experiment costs (Wahyono et al., 2019). At the same time, it prevents students from conducting dangerous science experiments using traditional methods, thereby creating a safer educational process. Indeed, in YZ, virtual situations not only provide students with practical experience but also allow them to try out and explore different scenarios. This improves the quality of education rather than merely increasing student participation in science education (Almasri, 2024). In the long term, AI-based simulations and virtual laboratories will make science education more accessible and comprehensive.

11.2.5 Redefining the Role of Teachers

AI teaching platforms have changed teachers' roles. Studies indicate that teachers do not fully leverage AI's great potential for learning and teaching (Celik, 2023). AI has a significant place in education and, therefore, closely affects teachers. AI education platforms do not completely eliminate teachers, but only lead to certain changes in their roles. These roles aim to have teachers become instructional designers, guides, and facilitators in these processes. In this regard, experts emphasize that no matter how advanced AI education systems become, they cannot replace teachers, and teachers will always be needed. Here, it is emphasized that AI will make teachers' jobs easier and enrich them with alternative options. Therefore, teachers must also learn AI educational technologies and be open to applying them in their classrooms, changing alongside them (Chiu et al., 2023; Straight, 2019).

11.3 Challenges Encountered in Artificial Intelligence Integration

11.3.1 Ethical and Privacy Concerns

It has been determined that AI cannot be addressed solely in terms of technology use by teachers; it has been emphasized that AI must also be addressed in terms of teachers' pedagogical, ethical, and teaching competence (Cooper & Tang, 2024;

Gouseti et al., 2024; Holmes et al., 2019). Therefore, certain ethical concerns are at the forefront of AI-based teaching systems. The continuous monitoring and analysis of student data through AI applications have raised issues of data security, ethical concerns, and the risk of misuse of private data. Consequently, protecting student privacy has been seen as the most important ethical issue in the long term (Lameras & Arnab, 2022). Another ethical issue encountered in AI applications is the production of "fake citations" (Gleason, 2022). This creates a data security problem. Therefore, to prevent these ethical issues, national and international institutions and organizations have published numerous AI ethical guidelines (Adams et al., 2023). At this stage, changes in teachers' evolving roles and curriculum updates within the framework of ethical principles for artificial intelligence should also be considered (Mai et al., 2024).

The ethical principles of AI include privacy, responsibility, freedom, autonomy, justice and fairness, transparency, and non-maleficence, and it is noted that these principles also apply to children. In addition, four new ethical principles have been established for K-12 education: pedagogical appropriateness, children's rights, artificial intelligence literacy, and teacher well-being (Adams et al., 2023). It has been observed that K-12 education lags in addressing and implementing AI ethics, but efforts are underway to remedy this in the near future (European Commission, Directorate-General for Education, Youth, Sport and Culture, 2022; The Institute for Ethical AI in Education, 2021).

11.3.2 Teacher Competencies and Education

The growing importance of AI in K-12 education has made it imperative that teachers have sufficient skills to effectively teach students using AI (Wang et al., 2023). However, studies have revealed gaps in the AI tool usage skills of K-12 teachers and students (Crompton et al., 2022). In addition, it has been found that teachers do not sufficiently utilize AI in teaching and learning (Celik, 2023). Therefore, the effective and efficient use of AI-based teaching systems depends on teachers receiving pedagogical training in these technologies and developing their technical skills. However, studies show that K-12 teachers lack sufficient knowledge and equipment to use AI for teaching. In addition, it has been found that some teachers do not approve of technology in the classroom and are resistant to change, preferring to teach using traditional methods. Teachers' concerns about using new techniques alongside evolving technology can dampen their efforts to use technology (Hébert et al., 2021). To resolve this situation, in-service training programs for K-12 teachers may need to be designed.

11.3.3 Inequality and Access Issues

The infrastructure and hardware necessary for AI educational technologies to be used everywhere and widely must be provided. Indeed, socio-economic differences can be an obstacle to this. Consequently, this situation contradicts the principle of equal opportunity in education. It is emphasized that caution must always be exercised regarding equality in AI applications (Cooper & Tang, 2024). Studies have shown that AI has great potential in education, but it can also pose challenges to educational equality and access. Indeed, it has been observed that systemic biases in AI algorithms and data can lead to inequalities (Adams et al., 2022). These biases are particularly related to race, gender, socioeconomic status, etc., and it is thought that they will produce unfair results that violate the principle of equality in the future (Zhai, 2022).

11.3.4 The Future Perspective of AI in Science Education

AI is not so much a groundbreaking tool in K-12 science education as it is an innovation that "transforms the nature of learning". The barriers to this innovation in science education must be removed, and this innovation must be sustainable. At this point, to ensure the sustainability of AI in science education:

- Including AI literacy in the relevant curriculum for training teacher candidates,
- Taking measures that will not cause ethical and privacy issues,
- It is necessary to ensure equality and access to technology, and, finally, to foster interdisciplinary collaboration.

The long-term future of AI in science education is expected to create a powerful "learning mechanism" in both cognitive and affective dimensions. Indeed, as AI continues to develop, its potential for integration into science education will also increase. This will provide students at all levels with a science education experience and enable them to transform these experiences (Almasri, 2024).

11.4 Conclusion

Considering the transformative impact of AI in K-12 science education, it will have long-term effects. This situation demonstrates significant potential for science education in personalized learning environments and assessment systems. However, for this great potential to be realized, certain problems must be addressed and necessary measures taken. Indeed, overcoming ethical, pedagogical, and technological barriers and ensuring the process is carried out in a healthy manner is imperative. Therefore, AI should not be viewed as a tool in the field of technology and science education, but rather as a mechanism that will shape the future of science education.

References

Adams, C., Pente, P., Lemermeyer, G., & Rockwell, G. (2023). Ethical principles for artificial intelligence in K-12 education. Computers and Education: Artificial Intelligence, 4, 100131. doi:https://doi.org/10.1016/j.caeai.2023.100131.

Adams, C., Pente, P., Lemermeyer, G., Turville, J., & Rockwell, G. (2022). Artificial intelligence and teachers' new ethical obligations. *The International Review of Information Ethics, 31*(1). https://doi.org/10.29173/irie483

Al Darayseh, A. (2023). Acceptance of artificial intelligence in teaching science: Science teachers' perspective. Computers and Education: Artificial Intelligence, 4, 100132. doi:https://doi.org/10.1016/j.caeai.2023.100132.

AlKanaan, H. M. N. (2022). Awareness regarding the implication of artificial intelligence in science education among pre-service science teachers. *International Journal of Instruction, 15*(3), 895–912. https://doi.org/10.29333/iji.2022.15348a

Almasri, F. (2024). Exploring the impact of Artificial Intelligence in teaching and learning of science: A systematic review of empirical research. *Research in Science Education, 54*, 977–997. https://doi.org/10.1007/s11165-024-10176-3

Asogwa, O., & Oladugba, A. (2015). Of students academic performance rates using artificial neural networks (ANNs). *American Journal of Applied Mathematics and Statistics, 3*(4), 151–155. https://pubs.sciepub.com/ajams/3/4/3

Baker, R. S., & Inventado, P. S. (2014). Educational data mining and learning analytics. *Learning Analytics*, 61–75.

Boulay, B. (2019). Escape from the skinner box: The case for contemporary intelligent learning environments. *British Journal of Educational Technology, 50*(6), 2902–2919. https://doi.org/10.1111/bjet.12860

Celik, I. (2023). Towards intelligent-TPACK: An empirical study on teachers' professional knowledge to ethically integrate artificial intelligence (AI)-based tools into education. Computers in Human Behavior, 138, 107468. doi:https://doi.org/10.1016/j.chb.2022.107468.

Chiu, T. K., Xia, Q., Zhou, X., Chai, C. S., & Cheng, M. (2023). Systematic literature review on opportunities, challenges, and future research recommendations of artificial intelligence in education. *Computers and Education: Artificial Intelligence, 4*, 100118. https://doi.org/10.1016/j.caeai.2022.100118

Cooper, G., & Tang, K. S. (2024). Pixels and pedagogy: Examining science education imagery by generative artificial intelligence. *Journal of Science Education Technoogy, 33*, 556–568. https://doi.org/10.1007/s10956-024-10104-0

Crompton, H., Jones, M. V., & Burke, D. (2022). Affordances and challenges of artificial intelligence in K-12 education: A systematic review. *Journal of Research on Technology in Education, 56*(3), 248–268. https://doi.org/10.1080/15391523.2022.2121344

D'Mello, S. K., & Graesser, A. (2015). Feeling, thinking, and computing with affect-aware learning technologies. In R. A. Calvo, S. K. D'Mello, J. Gratch, & A. Kappas (Eds.), *The Oxford handbook of affective computing*.

Demartini, C. G., Sciascia, L., Bosso, A., & Manuri, F. (2024). Artificial intelligence bringing improvements to adaptive learning in education: A case study. *Sustainability, 16*(3), 1347. https://doi.org/10.3390/su16031347

Deveci Topal, A., Dilek Eren, C., & Kolburan Geçer, A. (2021). Chatbot application in a 5th grade science course. *Education and Information Technologies, 26*(5), 6241–6265. https://doi.org/10.1007/s10639-021-10627-8

European Commission, Directorate-General for Education, Youth, Sport and Culture. (2022). *Ethical guidelines on the use of artificial intelligence (AI) and data in teaching and learning for educators.* Publications Office of the European Union. https://data.europa.eu/doi/10.2766/153756(open in a new window)

Gleason, N. (2022). ChatGPT and the rise of AI writers: How should higher education respond? *The Campus Learn, Share, Connect*. Retrieved January 13, 2023 from https://www.timeshigheducation.com/campus/chatgpt-and-rise-ai-writers-how-should-higher-education-respond

Gouseti, A., James, F., Fallin, L., & Burden, K. (2024). The ethics of using AI in K-12 education: A systematic literature review. *Technology, Pedagogy and Education, 34*(2), 161–182. https://doi.org/10.1080/1475939X.2024.2428601

Hébert, C., Jenson, J., & Terzopoulos, T. (2021). "Access to technology is the major challenge": Teacher perspectives on barriers to DGBL in K-12 classrooms. *E-Learning and Digital Media, 18*(3), 307–324. https://doi.org/10.1177/2042753021995315

Holmes, W., Bialik, M., & Fadel, C. (2019). *Artificial intelligence in education promises and implications for teaching and learning* (1st ed.). Center for Curriculum Redesign.

Hua Liu, H. B. (2024). Learning evaluation method based on artificial intelligence technology and its application in education. *Journal of Electrical Systems*. https://doi.org/10.52783/jes.1722

Hutson, J., Jeevanjee, T., Graaf, V., Lively, J., Weber, J., Weir, G., Arnone, K., Carnes, G., Vosevich, K., Plate, D., Leary, M., & Edele, S. (2022). Artificial intelligence and the disruption of higher education: Strategies for integrations across disciplines. *Creative Education, 13*, 3953–3980. https://doi.org/10.4236/ce.2022.1312253

Kamalov, F., Santandreu Calonge, D., & Gurrib, I. (2023). New era of artificial intelligence in education: Towards a sustainable multifaceted revolution. *Sustainability, 15*(16), 12451. https://doi.org/10.3390/su151612451

Karsenti, T. (2019). Artificial intelligence in education: The urgent need to prepare teachers for tomorrow's schools. *Formation et profession, 27*(1), 105–111. https://doi.org/10.18162/fp.2019.a166

Khan, I., Ahmad, A. R., Jabeur, N., & Mahdi, M. N. (2021). An artificial intelligence approach to monitor student performance and devise preventive measures. *Smart Learning Environments, 8*(1), 17. https://doi.org/10.1186/s40561-021-00161-y

Lameras, P., & Arnab, S. (2022). Power to the teachers: An exploratory review on artificial intelligence in education. *Information, 13*(1), 14. https://doi.org/10.3390/info13010014

Luckin, R., Holmes, W., Griffiths, M., & Forcier, L. B. (2016). *Intelligence unleashed: An argument for AI in education.* Pearson.

Mai, D. T. T., Van Da, C., & Van Hanh, N. (2024). The use of ChatGPT in teaching and learning: A systematic review through SWOT analysis approach. *Frontiers in Education, 9*. https://doi.org/10.3389/feduc.2024.1328769

Mavroudi, A., Giannakos, M., & Krogstie, J. (2018). Supporting adaptive learning pathways through the use of learning analytics: Developments, challenges and future opportunities. *Interactive Learning Environments, 26*(2), 206–220. https://doi.org/10.1080/10494820.2017.1292531

Ngo Ndjama, J. D. (2025). The use of artificial intelligence in lesson delivery and evaluation in large-scale teaching environments. *Interdisciplinary Journal of Education Research*. https://doi.org/10.38140/ijer-2025.vol7.s1.08

Owan, V. J., Abang, K. B., Idika, D. O., Etta, E. O., & Bassey, B. A. (2023). Exploring the potential of artificial intelligence tools in educational measurement and assessment. *Eurasia Journal of Mathematics, Science and Technology Education*. https://doi.org/10.29333/ejmste/13428

Romero, C., & Ventura, S. (2020). Educational data mining and learning analytics: An updated survey. WIREs Data Mining and Knowledge Discovery, 10(3), e1355. doi:https://doi.org/10.1002/widm.1355.

Shin, W.-S., & Shin, D.-H. (2020). A study on the application of artificial intelligence in elementary science education. Journal of Korean Elementary Science Education, 39(1), 117–132. doi.org/10.15267/KESES.2020.39.1.117.

Straight, N. (2019). Artificial intelligence in education: Uses and applications. *Block Letter*. https://robots.net/ai/artificial-intelligence-in-education-uses-and-applications/

Tapalova, O., Zhiyenbayeva, N., & Gura, D. (2022). Artificial intelligence in education: Aied for personalised learning pathways. *Electronic Journal of e-Learning*. https://doi.org/10.34190/ejel.20.5.2597

The Institute for Ethical AI in Education. (2021). The ethical framework for AI in education. *University of Buckingham*. https://www.buckingham.ac.uk/wp-content/uploads/2021/03/The-Institute-for-Ethical-AI-in-Education-The-Ethical-Framework-for-AI-in-Education.pdf

United Nations Educational, Scientific and Cultural Organization [UNESCO]. (2019). *The challenge and opportunities of artificial intelligence in education*. The United Nations Educational, Scientific and Cultural Organization.

Wahyono, I. D., Fadlika, I., Asfani, K., Putranto, H., Hammad, J., & Sunarti. (2019). New adaptive intelligence method for personalized adaptive laboratories. In *2019 International conference on electrical, electronics and information engineering (ICEEIE)* (pp. 196–200). https://doi.org/10.1109/ICEEIE47180.2019.8981477

Wang, S., Sun, Z., & Chen, Y. (2023). Effects of higher education institutes' artificial intelligence capability on students' self-efficacy, creativity and learning performance. *Education and Information Technologies, 28*(5), 4919–4939. https://doi.org/10.1007/s10639-022-11338-4

Yang, T.-Y., Brinton, C. G., Joe-Wong, C., & Chiang, M. (2017). Behavior-based grade prediction for MOOCs via time series neural networks. *IEEE Journal of Selected Topics in Signal Processing, 11*(5), 716–728. https://doi.org/10.1109/JSTSP.2017.2700227

Zhai, X. (2022). ChatGPT user experience: Implications for education. doi:https://doi.org/10.2139/ssrn.4312418

Zhai, X., Shi, L., & Nehm, R. H. (2021). A meta-analysis of machine learning-based science assessments: Factors impacting machine-human score agreements. *Journal of Science Education and Technology, 30*(3), 361–379. https://doi.org/10.1007/s10956-020-09875-z

Zhao, L., Chen, L., Liu, Q., Zhang, M., & Copland, H. (2019). RETRACTED: Artificial intelligence-based platform for online teaching management systems. *Journal of Intelligent & Fuzzy Systems: Applications in Engineering and Technology., 37*(1), 45–51. https://doi.org/10.3233/JIFS-179062

Chapter 12
AI Integration in Math, Science and Information Class

Benjamin Oh and Cheol Hong Jeon

12.1 Introduction

Integrating AI into STEM subjects aligns with a broader shift toward inquiry-based and interdisciplinary learning (Zhai & Krajcik, 2024). For instance, science classes use AI to analyze large data sets, simulate complex phenomena, and categorize data using machine learning algorithms (Tan et al., 2022; Zha et al., 2025; Zhai & Krajcik, 2024). In mathematics and information technology classes, students learn programming languages and employ LLMs to assist with problem solving and coding (Malik et al., 2025). Many researchers and educators hope these innovations will create opportunities for more personalized and engaging learning experiences (Lin, 2022; Malik et al., 2025; Park et al., 2023; Shi et al., 2024; Tan et al., 2022; Touretzky et al., 2019). Essentially, they simulate a "more knowledgeable other" in Vygotsky's zone of proximal development, which provides efficient scaffolding (Vygotsky & Cole, 1978; Wood et al., 1976). In other words, generative AI like ChatGPT can provide dialogue-based hints or suggest next steps in real time, much like a peer tutor would, thereby extending a student's capabilities (Tan et al., 2022; Zha et al., 2025; Zhai & Krajcik, 2024). However, some caution that AI can hinder learning if students depend on it for answers without developing understanding or if the AI's guidance causes confusion (Park et al., 2023; Shi et al., 2024; Zha et al., 2025). Moreover, using

The original version of the chapter has been revised. A correction to this chapter can be found at https://doi.org/10.1007/978-3-032-18030-8_14

B. Oh
Lynbrook High School, San Jose, CA, USA
e-mail: benoh893@gmail.com

C. H. Jeon (✉)
Korea National University of Education, Cheongju, Korea
e-mail: ferrate87@knue.ac.kr

© The Author(s), under exclusive license to Springer Nature Switzerland AG 2026, corrected publication 2026

M. Alkış Küçükaydın (ed.), *Preparing K-12 Pre-Service Science Teachers for the Age of Artificial Intelligence*, Advances in Technology-Rich Science Education 4,
https://doi.org/10.1007/978-3-032-18030-8_12

AI in subject areas requires digital infrastructure and basic AI literacy skills, as well as a proper curriculum that effectively integrates subject content and AI (Zhai & Krajcik, 2024). Thus, we present how AI integration — the scaffolds provided, and the culture of use established — determines whether AI empowers students or undermines their learning. To fully understand this dynamic, we also need to consider the student's internal experience, which is not purely cognitive. Therefore, we draw upon recent literature that underscores the interwoven relationship between cognition and emotion to provide a more holistic framework for our analysis.

In this chapter, we present three case studies. The first case study examines high school science classrooms in Korea and the United States, where AI tools were integrated into inquiry-based projects. The study reveals how students in different cultural contexts respond cognitively and emotionally to AI tools. Case 2 involves a team of university instructors from various disciplines (e.g., data science, computer science, and statistics) who collaborated to incorporate LLMs into an undergraduate data science course. This case study illustrates the considerations involved in using AI as a scaffolding tool. Case 3 takes place in an innovative chemistry lab where pre-service teachers use physical computing devices, such as Arduino sensors, and ChatGPT to conduct experiments. This case study highlights how students' problem-solving approaches and AI literacy evolve over time. Through analyzing these cases, we illustrate AI's dual role as a helpful tool and a potential obstacle, which can help researchers and educators design effective, AI-integrated instruction.

AI functions as both a scaffold and a barrier to learning. As a scaffold, AI offers conceptual and procedural support. It does so by explaining concepts, guiding problem-solving steps, and providing adaptive feedback (Liu & Adams, 2017; Vygotsky & Cole, 1978). Generative AI can also provide emotional support by encouraging learners and alerting teachers when learners become frustrated (Kohnke & Moorhouse, 2025). Systems such as Inq-ITS and tutoring AIs diagnose learners' needs and personalize support, thereby enhancing engagement and cognitive growth (Gobert et al., 2024; Malik et al., 2025; Zhai & Krajcik, 2024). However, AI can act as a barrier when it fosters dependence, misinformation, or cognitive overload, particularly in the absence of proper infrastructure or teacher guidance (Park et al., 2023; Shi et al., 2024; Zha et al., 2025). Whether AI supports or hinders learning depends on effective mediation and AI literacy among teachers and students.

12.2 Background

AI-mediated inquiry also reshapes emotional dynamics. Epistemic emotions, such as curiosity, confusion, frustration, and anxiety, strongly influence engagement (Oh & Han, 2021; Pekrun, 2006). Positive emotions, such as curiosity, enhance exploration. Negative emotions, if managed effectively, can also deepen understanding (Chen et al., 2019; Jaber & Hammer, 2016; Tomas et al., 2016). An effective AI-integrated pedagogy must therefore account for this full spectrum of epistemic emotions. Teachers should monitor students' emotional responses alongside their cognitive progress, intervening when necessary to help students harness their emotions constructively. For instance, a brief confusion can be reframed as a signal to investigate further, and a bout of frustration can be mitigated by encouragement or hinting at a strategy, turning it into a learning moment. By treating emotions like curiosity, confusion, and even frustration as *resources for learning* rather than simply as symptoms to eliminate, educators can create a more resilient learning environment (Jaber & Hammer, 2016; Pekrun, 2006). This theoretical perspective will inform our analysis of the cases, as we pay special attention to how AI as scaffold or barrier is often mediated by the emotional context and how teachers and students respond to the feelings that arise during AI-supported learning.

12.3 Case Study 1: Comparative Study of Korea–U.S. High School Science Classrooms

The first case study compares the integration of artificial intelligence (AI) into inquiry-based learning in high school science classrooms in South Korea and the United States. Through text mining of classroom discourse and student survey data, the study examined the cognitive and emotional engagement of teachers and students with AI-supported activities, such as data modeling, simulation, and problem solving. The analysis revealed contrasting instructional patterns shaped by cultural norms and teaching styles. This chapter interprets these findings through the lens of cognitive and emotional scaffolding, highlighting how AI functioned as both a learning support and a source of new challenges in each context.

12.4 Cognitive Scaffolding

In South Korea, AI was woven into structured science tasks as a cognitive scaffold to reinforce concepts. Teachers posed well-defined scientific problems (e.g. analyzing long-term climate change data on the Korean Peninsula) and students used AI tools to find correct solutions (Ministry of Education, 2022; Han et al., 2023). Class activities often focused on data modeling and big-data analysis of environmental issues (e.g., climate trends or species life cycles), emphasizing conceptual understanding. Korean students discussed how AI works and its scientific implications, blending computing with science content. Survey results reflect this focus: Korean students reported higher sustained engagement and greater learning gains in data analysis and problem-solving than their U.S. peers. However, this rigorous, teacher-guided approach meant students also felt a heavier cognitive load.

In the United States, AI served as a scaffold for exploration and hands-on problem-solving. Lessons were more open-ended – for example, using AI for Oceans (a machine learning activity classifying fish and marine debris) or wildfire prediction projects to learn by doing(Touretzky et al., 2019). The pedagogy favored interactive collaboration over fixed outcomes. Students and teachers engaged in constant dialogue ("Guys, let's try...") to troubleshoot and iterate on AI-driven tasks(Alsop & Watts, 2003). Topic modeling confirms U.S. classroom discourse was filled with informal team-based talk and technical terms related to debugging and moving robots. This indicates the cognitive scaffolding was distributed: peers and AI tools together guided learning. U.S. students highly valued peer collaboration (per surveys), though they rated teacher explanations and clarity lower than Korean students did. In sum, U.S. classrooms scaffolded learning through collaborative experimentation, which encouraged practical skills but sometimes lacked the explicit conceptual framing seen in Korea.

In comparison, both countries integrated AI into science inquiry and often tackled environmental topics. The difference lay in instructional scaffolding: Korean classes provided more structured, concept-heavy guidance, whereas U.S. classes provided more peer-driven and improvisational support. Each approach yielded cognitive engagement but of different flavors – one driven by mastery of content, the other by iterative problem-solving.

12.5 Emotional Scaffolding

In South Korea, the structured approach came with intense emotional investment. Korean students frequently expressed excitement at successes ("It was so exciting!") and pride in learning new AI skills. Teachers acknowledged achievements to keep morale high. However, negative emotions were more common in the Korean class than the American class (~10.5% vs 6.0% of utterances were negative). This higher percentage of negative utterances does not necessarily indicate a worse learning experience; rather, it could reflect a classroom culture where expressing struggle is permissible and immediately addressed. Frustration often arose when facing technical hurdles or complex data. Notably, students didn't give up; they treated setbacks as puzzles to be solved ("We found the error and we will fix it"). This suggests that, despite anxiety, the classroom culture (and teacher support) encouraged perseverance. Emotional scaffolding in Korea involved the teacher providing reassurance and step-by-step feedback. By celebrating completed tasks and guiding students through challenges, teachers helped students channel frustration into determination. Still, survey data show Korean students felt a heavier workload and time pressure, indicating stress that needs careful management through supportive feedback and pacing.

The U.S classes provide an implicit emotional scaffold: a collaborative, open atmosphere where challenges are "no big deal." Teacher intervention focused less on direct instruction and more on keeping the group motivated and on task. The emotional climate was positive and relatively low-stress. Teachers continuously encouraged students to readily celebrate specific small achievements, which helped the class maintain enthusiasm. Importantly, overt frustration was rarer in U.S. classrooms. When issues arose (glitches, confusing concepts), students tended to discuss them matter-of-factly or with humor rather than as personal failings. The result was that technical problems were viewed as practical obstacles, not emotional setbacks. Survey responses reflect this: U.S. students reported slightly less anxiety about workload and felt technical issues were manageable. In short, the U.S. approach scaffolded positive emotions by normalizing trial-and-error and providing peer support, making the learning process enjoyable and resilient to stress.

Both settings showed high engagement and confidence using AI, but emotional scaffolding differed. Korean classes, with their high standards, required teachers to continually coach students through anxiety and celebrate successes to keep students motivated. American classes relied on a light-hearted, team-based ethos to sustain confidence, needing less direct emotional intervention. Each approach has trade-offs: Korean students developed a strong sense of accomplishment but felt more pressure, whereas U.S. students had more fun and peer encouragement but sometimes at the expense of clear guidance.

12.6 Epistemic Emotions

Although definitions of epistemic emotions vary by field, a shared concept is that these are emotions experienced when engaging in epistemic activities, such as acquiring, creating, and evaluating knowledge. Representative epistemic feelings include curiosity, frustration, excitement, surprise, and confusion.

In South Korea, curiosity and confusion—key epistemic emotions related to learning—were strongly evident. Korean students showed intellectual curiosity, often wondering how AI works or what they could discover ("I'm so curious about this ability…"). The structured tasks (like climate-data modeling) sparked genuine interest in scientific inquiry with AI. At the same time, confusion and uncertainty were more openly expressed in the Korean class. Students voiced when they felt overwhelmed by a complex dataset or an unclear AI result, and they openly asked for help or clarification. This direct expression of confusion can be productive: it allowed teachers to step in with explanations or hints, and it often led students to collaboratively reason through problems. Indeed, Korean discussions even included debating AI's decisions and ethics, indicating students were grappling with understanding—an intellectual struggle accompanied by epistemic emotions like uncertainty or doubt. The challenge for teachers was to harness these emotions: turning confusion into a learning opportunity (through guided debate or problem-solving) and sustaining curiosity despite difficulties. When successful, this led to deep engagement; when not, it risked anxiety. The higher anxiety noted in Korean students suggests that epistemic uncertainty (not knowing immediately how to proceed) was stressful, underscoring the need for scaffolding strategies that build resilience (e.g. breaking tasks into smaller steps, providing timely hints).

U.S. students also experienced epistemic emotions, but these manifested differently. Curiosity was evident in their eagerness to experiment – they enjoyed exploring new AI tools and seeing what they could create or improve (for instance, tweaking a model to boost accuracy or trying a creative solution in the AI for Oceans activity). This hands-on curiosity kept them engaged. Confusion and doubt occurred as well – AI projects inevitably presented concepts that students found complex or "fancy" – but American students were less likely to label their feelings as confusion openly. Instead, they often externalized it: a student might remark on a confusing concept by saying "that's just a fancy way of saying X," framing it as an observation rather than a personal lack of understanding. Likewise, challenges (like a "glitchy" tool or a tough algorithm) were discussed in technical terms rather than emotional terms. This suggests that the classroom norm encouraged maintaining a can-do attitude, students implicitly treated confusion as a normal part of tinkering. The benefit is that they remained optimistic and willing to continue experimenting without fear of failure. However, one risk is that hidden confusion might go unaddressed – since students don't always explicitly ask for help, teachers need to infer when deeper conceptual support is needed. Survey findings support that U.S. students were generally less worried about the AI's complexity but did occasionally note difficulty understanding results. Effective scaffolding in the U.S. context involved encouraging questions and reflection so that latent confusion could surface and be resolved.

In both countries, *epistemic emotions* like curiosity, confusion, and surprise played a pivotal role in the AI learning process. Both Korean and American students were highly curious and took pride in their AI-driven discoveries, indicating that AI can ignite intellectual excitement across cultures. Both also encountered confusion and uncertainty – a natural byproduct of learning cutting-edge content – and needed support to navigate these feelings. The difference lies in expression and management: Korean students voiced curiosity and confusion more explicitly, allowing direct instructional scaffolding but also experiencing more anxiety when facing the unknown. American students, buoyed by a collaborative maker culture, expressed curiosity through active experimentation and downplayed confusion, which kept anxiety low but put more onus on teachers to detect who might be quietly puzzled. For educators, these insights highlight the importance of designing AI activities that not only challenge students cognitively but also support them emotionally. Fostering a classroom culture that treats curiosity as the driving force and normalizes confusion as *part of learning* – while providing guidance to resolve that confusion – is key. Both approaches underscore that managing epistemic emotions is crucial: with proper scaffolding, initial confusion can lead to deeper inquiry and learning, and curiosity can flourish into lasting interest in AI and science.

12.7 Case Study 2: Data Science Instructor Collaboration to Integrate LLMs

The second case illustrates an instructor collaboration group among introductory data science instructors in the Data Science and AI Academy (DSA) at North Carolina State University. Since 2021, DSA has been offering introductory project-based data science courses for all students across the university without prerequisites. However, these efforts to broaden access to data science education were faced with complex problems of practice, which pushed the instructor community to design and implement instructional innovations that can better support student learning of data science in relevant ways. One of the key problems of practice has been making the code-first curriculum accessible to students with diverse disciplinary backgrounds, many of whom have had limited programming backgrounds.

12.8 Emergence of AI Integration as a Common Problem of Practice

In the early years of DSA (2021–2024), instructors started to notice the emergence of a type of generative AI, Large Language Models (LLMs, e.g., ChatGPT, Gemini). Instructors approached the use of LLMs in their courses in a variety of ways. Some instructors allowed students to freely use LLMs, while others entirely banned LLMs due to concerns about students' overreliance. As years passed by, instructors started noticing the increased student use of LLMs, seeing student submissions including LLM hallucination (inaccurate but seemingly plausible responses). These concerns about students' misuse of AI tools motivated instructors to develop a common pedagogical approach, through which the instructors can guide students to effectively and ethically use AI tools for their data science learning.

Informed by the TPACK framework (Koehler & Mishra, 2009), the instructor collaboration began with the goal of integrating technological knowledge (i.e., knowledge of LLMs) with pedagogical knowledge (i.e., knowledge of teaching and learning) and content knowledge (i.e., disciplinary knowledge of data science). In Fall 2024, the collaborative group began, which consisted of seven participants with complementary backgrounds, such as data science education, teacher education, computer science education, machine learning and AI, and industry practice. The collaboration led to the development of a pedagogical approach named *LLM prompt templates* (Byun et al., 2025). During the collaboration and knowledge integration, nuanced insights on AI integration in introductory data science courses emerged, and we present some of the relevant insights below.

12.9 Core Issue with Transferring Industry Use of LLMs into the Classroom

The key motivation for integrating LLMs in data science courses was instructors' experiences with the changing data science industry. It became a norm for data scientists to use LLMs effectively to increase their time efficiency on their tasks. Entry-level programming tasks have been automated and replaced by LLMs; therefore, data science instructors felt responsible for preparing students for the changing data science landscape and newly emerging knowledge and practices necessary for working with LLMs. Some instructors who had previously banned the use of LLMs began to shift toward including LLMs in their courses to be responsive to these industry needs.

The collaborative process, however, highlighted that the industry use of LLMs cannot be easily transferred to classrooms. Although the use of LLMs in the data science industry and the classroom may appear similar, there are distinctive purposes of LLM use. In the industry setting, LLMs are used to improve time efficiency for knowledgeable users who can discern the validity of LLM output. On the other hand, in the classroom setting, LLMs are used to engage students with cognitive processes to learn data science concepts and skills. That is, the *disciplinary use of* LLMs cannot be equated with the *disciplinary learning with* LLMs. Addressing this difference in LLMs in industry and classroom settings became the primary consideration of the instructor collaboration.

12.10 LLM Prompt Template to Highlight the Scaffolding Purpose of LLMs

To make the transition from the use of LLMs to learning with LLMs, the collaborative group developed the pedagogical method of the LLM prompt template. Templates can be designed by instructors for students to prompt LLMs. Prompt templates include both fixed and modifiable components (i.e., fill-in-the-blank slots, see Chang et al., 2023). The fixed components offer a structure of student prompting guided by the instructor's pedagogical purpose of scaffolding, but the modifiable components allow students to fit the prompt in the specific context of their learning tasks. Here is an example of a prompt template (Byun et al., 2025):

> GENERATE code with **[R tidyverse/Python pandas]** to **[a type of wrangling step]** for the dataset **[Dataset Name (object name not datafile name)]** which includes variables **[list of key variable]**, and briefly EXPLAIN step-by-step in plain language how this code works.

The example above is designed based on the pedagogical purpose of scaffolding students in generating R or Python code for a specific step of data wrangling (e.g., filter, mutate). The modifiable components are marked with bold text, which allows students to modify the prompt to get help for their own specific problems or tasks.

Offering this prompt template to students can encourage their LLM use for the specified pedagogical purpose, while the template functions as a guide. Without these types of structure, novice students often fall into using vaguely constructed prompts that do not give adequate context for LLMs to generate useful output. In extreme cases, students may prompt the entire assignment in an attempt to gain useful output from LLMs. However, LLM outputs may not provide relevant details or underlying understandings for students to critically examine and learn from.

12.11 Use of Prompting Techniques for Enhancing Scaffolding

As the instructor collaboration progressed, instructors started to view prompting techniques (technological knowledge) through a lens of teaching and learning (pedagogical knowledge). This integrated knowledge is also encoded in prompt templates as a key feature. For instance, chain-of-thought prompting (Wei et al., 2023) is a technique to facilitate particular reasoning steps by LLMs. By adding "EXPLAIN step-by-step …" in the above prompt template, student prompts can elicit reasoning steps from the model, which can offer "an interpretable window into the behavior of the model" (p. 3). In other words, the chain-of-thought prompting technique can be used to offer opportunities for students to think through the LLM output and better serve the pedagogical purpose of scaffolding.

In sum, the knowledge integration among instructors, encapsulated as the LLM prompt template, suggests that the use of AI for scaffolding is not a trivial matter. As digital technologies like LLMs can be used in a wide range of ways, instructors need to judiciously introduce the use of AI tools that are fit for the intended pedagogical purposes. The prompt template exemplifies such efforts to narrow the use of AI tools, allowing students to engage with and benefit from AI tools while mitigating the risk of overreliance and ineffective use.

12.12 Case Study 3: AI-Mediated Chemistry Inquiry as Scaffolding for Pre-Service Teachers Integration of AI in an Open-Ended Lab.

In this case, a university chemistry education program redesigned its lab course for pre-service science teachers to include AI as a scaffolding tool. Nineteen first-year graduate students (future secondary school chemistry teachers) participated in open-ended inquiry projects, using Arduino-based sensors and data loggers alongside ChatGPT (a large language model) as a virtual assistant. Each team formulated a research question, designed an experiment with the sensors, collected and analyzed data (often coding with Python for sensor data), and drew conclusions—mirroring real scientific inquiry. ChatGPT was available at every stage: from brainstorming hypotheses and troubleshooting equipment to getting coding help and interpreting results. There was no single "right answer" in these inquiries; success depended on how well students iterated on their experiments and refined their understanding, with ChatGPT acting as a collaborative support. This design (part of a larger research project) aimed to modernize chemistry labs with digital tools and also foster AI literacy among future teachers, giving them first-hand experience with AI-supported learning.

12.13 ChatGPT's Role as a Scaffold in Inquiry

Much like a versatile teaching assistant, ChatGPT functioned as a multifaceted cognitive scaffold that students could consult for various needs, primarily for help with programming tasks. Since experiments involved custom code for Arduino sensors and data analysis, students often asked ChatGPT to help them write or debug code. For instance, a team might inquire, "How do we program the Arduino to record the temperature every second and save the data?" ChatGPT would then generate sample code or outline the steps, effectively scaffolding the technical aspect of the inquiry. Roughly one-quarter of students' questions to ChatGPT were basic code generation requests, and ~ 14% were debugging queries. This highlights the AI's crucial role in the technical execution of experiments. This support lowered the barrier for students with limited programming experience, allowing them to focus on scientific questions while gradually learning to code through the AI's guidance. ChatGPT also provided conceptual scaffolding for the inquiry process. Students asked theoretical or design-related questions, such as "What factors might affect the pH change in this reaction?" and "How can we measure reaction rate with these sensors?" In response, ChatGPT suggested variables to consider, possible experimental setups, and relevant scientific concepts. Though these conceptual questions were less common than requests for coding assistance, some teams leveraged ChatGPT to explore underlying chemical theories or alternative strategies when an approach failed. In this way, ChatGPT expanded students' thinking by quickly supplying background knowledge and ideas. It served as a "more knowledgeable other" that could bridge gaps in understanding and help them frame their experiments.

After collecting data, students often turned to ChatGPT for help interpreting the results. They would summarize their observations, share a pattern (e.g., "Our temperature readings keep dropping irregularly over time"), and ask for possible explanations or analysis methods. Although ChatGPT could not see the raw data, students learned to describe data trends and receive input from the AI on what those trends might mean. For example, ChatGPT might suggest, "Perhaps heat loss to the environment could cause that drop. You could try fitting an exponential decay curve." This kind of guidance acted as a thinking partner, prompting students to consider analysis techniques or explanations they might not have thought of on their own. The interactions were often dialogic; students would go back and forth with the AI, refining their questions and asking for clarification, much as they would with a human tutor. Some students even asked ChatGPT to critique their hypotheses or conclusions (e.g., "We think the inconsistent pH readings are due to sensor calibration drift — does that make sense?"). In these moments, the AI served as a sounding board, providing feedback on their reasoning. This immediate, conversational support propelled students forward in their investigations, demonstrating how AI can facilitate inquiry by offering timely guidance and expert-level insights on demand.

12.14 Epistemic Emotions

Throughout the AI-supported chemistry inquiry, students experienced a range of epistemic emotions, including curiosity, confusion, frustration, and satisfaction. These emotions played pivotal roles in the students' learning process. Initially, many pre-service teachers approached the open-ended lab with genuine curiosity and excitement. The novelty of using ChatGPT to design experiments and troubleshoot code sparked their eager exploration. For example, students brainstormed unique questions, such as, "What if we vary the water volume to see different boiling behaviors?" They were eager to see how the AI might help answer these questions. This initial curiosity was a driving force that kept students engaged. This is consistent with research showing that positive epistemic emotions, such as curiosity, can motivate deeper inquiry and persistence in complex tasks. The AI tool further fueled this curiosity by providing immediate answers and new ideas, which reinforced students' enthusiasm to try multiple approaches. For example, one student exclaimed, "ChatGPT gave us a totally new idea to test!" This early encouragement set an optimistic tone for the inquiry phase.

As the investigation progressed, confusion naturally arose when theory met practice. Students often encountered uncertainties when interpreting ChatGPT's guidance or making sense of unexpected experimental results. In one instance, a team was perplexed by a diagrammatic wiring explanation from ChatGPT and exclaimed, "This is so confusing! What does it want us to do?" The AI's pictorial instructions for connecting an Arduino sensor were unclear at first. The classroom teacher then stepped in, rephrasing the AI's advice and physically demonstrating the wiring. This support acted as a crucial cognitive scaffold, helping students link the AI's abstract explanation to the actual circuit on their breadboard and resolving confusion. According to scaffolding theory, timely intervention by a more knowledgeable person (in this case, the teacher) helps learners stay within their zone of proximal development — the sweet spot where challenges are manageable with guidance. Peers also played a role in regulating confusion. Students shared their perplexities with each other ("Do you understand what it means?"), and sometimes, a classmate who had solved a similar issue would offer a tip or some reassurance. These social and instructional supports prevented confusion from leading to discouragement. Instead, they framed confusion as a temporary hurdle on the path to understanding.

Frustration mainly surfaced during debugging cycles or when ChatGPT produced inaccurate answers. When the code repeatedly failed, some students expressed exasperation ("Why isn't this working?"), yet they persisted with AI and peer guidance until they succeeded. Teachers' emotional scaffolding, such as calm reminders to take breaks or reframe the task, helped students channel their frustration into problem solving. For instance, after ChatGPT incorrectly identified a chemistry equation, the students became skeptical and said, "That can't be right. Let's double-check." This moment of skepticism evolved into a critical learning opportunity that reinforced the importance of verification.

Moments of satisfaction punctuated the inquiry process when students finally resolved errors or obtained reliable data. Shouts of "It finally works!" captured the emotional payoff of sustained effort. Teachers reinforced these successes by prompting reflection, asking what strategies led to breakthroughs, thereby converting satisfaction into confidence and metacognitive awareness. Over time, repeated cycles of struggle and resolution helped students internalize emotional regulation. They learned that confusion signals curiosity, frustration can be productive, and satisfaction validates persistence.

By the end of the course, most pre-service teachers had shifted from relying too heavily on ChatGPT to using it critically and strategically. They recognized when to verify information or reason independently. This transition demonstrates that scaffolding epistemic emotions through timely support, dialogue, and reflection enabled learners to not only master AI-assisted inquiry, but also develop emotional resilience and confidence in tackling complex and uncertain problems.

12.15 Discussion

A core finding that cuts across these case studies is that AI played a dual role in the STEM classroom: it was both a helpful scaffold and a potential barrier. As a powerful tool, AI offered new methods for visualizing data, generating ideas, and receiving personalized feedback, thereby stimulating student inquiry and curiosity. However, technical glitches, opaque reasoning, and "hallucinated" outputs often led to frustration, confusion, and, at times, cognitive overload.

In this environment of dual-natured AI, students commonly demonstrated a developmental pattern of moving from dependence to autonomy. Initially, they relied heavily on AI to generate answers. However, with experience and guidance, they gradually transitioned to a more critical stage, using AI as a "collaborative partner" to verify their own thinking and explore alternatives. This suggests a "fading scaffold" effect, where AI support is internalized over time. This shows that when managed well, AI does not create long-term dependency, but instead helps students grow into more independent and critical thinkers.

The positive transformations observed in students—their growth in autonomy and ability to overcome frustration—did not occur automatically. Across all cases, the most critical variable in determining the educational success of AI was the teacher.

By acting as a "mediator" between students and the AI, teachers played a pivotal role in determining whether the AI would become a catalyst for learning or an obstacle.

In successful classrooms, instructors did more than just introduce technology. They: They provided nuanced guidance to alleviate anxiety about AI's unreliability and focus students on conceptual understanding (Korean case). They prompted reflection and discussion to ensure exploration with AI translates into deep learning rather than superficial play (U.S. case). They established norms for critically vetting AI outputs and intervened to turn errors into learning opportunities. Thus, the teacher's role expanded from content delivery to serving as a "coach for human-AI interaction," guiding students on how to question AI, cope with frustration, and use the technology productively. Ultimately, this "human scaffolding" from well-prepared teachers was key to harnessing AI's potential while mitigating its downsides.

These findings point to three specific recommendations for designing AI-integrated education. First, intentionally design for "fading scaffolds." AI should be used as a tool to extend, not replace, student thinking. Initially, educators can provide structured prompts or step-by-step guides to help students use AI to tackle difficult tasks. As students become more confident, learning activities should shift to require more independent decision-making and critical thinking, positioning AI as a consultant rather than an answer machine. The goal is to guide students from "asking AI for the answer" to "using AI to improve my answer." By assessing the process and reasoning, not just the final output, educators can prevent the superficial use of AI.

Second, intentionally incorporate emotional scaffolding. Learning with AI evokes strong "epistemic emotions," such as confusion and frustration. Instructional design should treat these emotional fluctuations as a normal part of the learning process. Teachers can normalize struggle by telling students, "The AI might give you strange answers, and figuring out why is part of the learning process." When frustration arises, instructors should model calm, systematic coping strategies. Incorporating brief reflection activities in which students discuss their emotional experiences can build resilience and help them view challenges as shared learning opportunities rather than personal failures.

Third, actively invest in teacher professional development. In the age of AI, the teacher's role becomes more critical, not less. Institutions must provide hands-on training that allows educators to experience the same AI tools as their students so that educators can understand the tools' capabilities and limitations. This training should cover concrete pedagogical strategies, such as how to establish guidelines for AI usage, troubleshoot common issues, and intervene appropriately (e.g., by asking, "Why do you trust that result?") to encourage critical thinking. When teachers feel confident and prepared, AI can become a powerful classroom tool that complements their support rather than replacing it.

AI is more than a simple technological tool. It is a complex variable that reshapes classroom culture, emotional dynamics, and pedagogy itself. This research makes it clear that AI's educational value is not inherent in the technology itself. Rather, it is realized through human-centered instructional design that purposefully builds student autonomy and carefully supports the emotional dimension of learning. This design is facilitated by well-prepared teachers. Therefore, the future of education should focus less on making AI smarter and more on wisely orchestrating learning environments where students can learn more deeply with AI and grow into adaptable, critical thinkers.

References

Alsop, S., & Watts, M. (2003). Science education and affect. *International Journal of Science Education, 25*(9), 1043–1057.

Byun, S., McClure, J., Ferrell, M., Faith, J., Zakaria, Z., Porter, K. E., & Harding, M. (2025). *Designing LLM prompt templates: A self-study of instructor collaboration and knowledge integration* [Manuscript submitted for publication]

Chang, M., Druga, S., Fiannaca, A. J., Vergani, P., Kulkarni, C., Cai, C. J., & Terry, M. (2023, June). The prompt artists. In *Proceedings of the 15th conference on creativity and cognition* (pp. 75–87).

Chen, Y.-C., Benus, M. J., & Hernandez, J. (2019). Managing uncertainty in scientific argumentation. *Science Education, 103*(5), 1235–1276.

Education, M. of. (2022). *Informatics and AI education in the 2022 revised national curriculum (in Korean).* Ministry of Education, South Korea.

Gobert, J. D., Li, H., Dickler, R., & Lott, C. (2024). Can AI-based scaffolding promote students' robust learning of authentic science practices. In X. Zhai & J. Krajcik (Eds.), *Uses of artificial intelligence in STEM education* (pp. 250–269). Oxford University Press.

Han, S., Lee, H., & Kim, J. (2023). Teachers' readiness for AI-powered science teaching: A comparative study of South Korea and the US. *Journal of Science Education Research, 32*(1), 45–60.

Jaber, L. Z., & Hammer, D. (2016). Learning to feel like a scientist. *Science Education, 100*(2), 189–220.

Koehler, M., & Mishra, P. (2009). What is technological pedagogical content knowledge (TPACK)? *Contemporary Issues in Technology and Teacher Education, 9*(1), 60–70.

Kohnke, L., & Moorhouse, B. L. (2025). Enhancing the emotional aspects of language education through generative artificial intelligence (GenAI): A qualitative investigation. *Computers in Human Behavior, 167*, Article 108600. https://doi.org/10.1016/j.chb.2025.108600

Lin, H. (2022). Influences of artificial intelligence in education on teaching effectiveness: The mediating effect of teachers' perceptions of educational technology. *International Journal of Emerging Technologies in Learning (Online), 17*(24), 144–156. https://doi.org/10.3991/ijet.v17i24.36037

Liu, J. C., & Adams, A. (2017). *Design of online student orientation with conceptual and procedural scaffolding BT - learning and knowledge analytics in open education: Selected readings from the AECT-LKAOE 2015 summer international research symposium* (F.-Q. Lai & J. D. Lehman (eds.); pp. 41–68). Springer International Publishing. https://doi.org/10.1007/978-3-319-38956-1_5

Malik, R., Abdi, D., Wang, R., & Demszky, D. (2025). Scaffolding middle school mathematics curricula with large language models. *British Journal of Educational Technology, 56*(3), 999–1027. https://doi.org/10.1111/bjet.13571

Muis, K. R., Chevrier, M., & Singh, C. A. (2018). The role of epistemic emotions in personal epistemology and self-regulated learning. *Educational Psychologist, 53*(3), 165–184.

Oh, P. S., & Han, M. (2021). A review of the history and recent trends of emotion research in science education. *Journal of the Korean Association for Science Education, 41*(2), 103–114.

Park, J., Teo, T. W., Teo, A., Chang, J., Huang, J. S., & Koo, S. (2023). Integrating artificial intelligence into science lessons: Teachers' experiences and views. *International Journal of STEM Education, 10*(1), 61. https://doi.org/10.1186/s40594-023-00454-3

Pekrun, R. (2006). The control–value theory of achievement emotions: Assumptions, corollaries, and implications for educational research and practice. *Educational Psychology Review, 18*(4), 315–341.

Pekrun, R., Goetz, T., Titz, W., & Perry, R. P. (2002). Academic emotions in students' self-regulated learning and achievement: A program of qualitative and quantitative research. *Educational Psychologist, 37*(2), 91–105.

Pekrun, R., Vogl, E., Muis, K. R., & Sinatra, G. M. (2017). Measuring emotions during epistemic activities: The Epistemically-Related Emotion Scales. *Cognition and Emotion, 31*(6), 1268–1276.

Shi, L., Ding, A.-C. (., & Choi, I. (2024). Investigating Teachers' use of an AI-enabled system and their perceptions of AI integration in science classrooms: A case study. In *Education Sciences, 14*(11). https://doi.org/10.3390/educsci14111187

Tan, S. C., Lee, A. V. Y., & Lee, M. (2022). A systematic review of artificial intelligence techniques for collaborative learning over the past two decades. *Computers and Education: Artificial Intelligence, 3*, Article 100097. https://doi.org/10.1016/j.caeai.2022.100097

Tomas, L., Rigano, D., & Ritchie, S. M. (2016). Students' regulation of their emotions in a science classroom. *Journal of Research in Science Teaching, 53*(2), 234–260.

Touretzky, D. S., Gardner-McCune, C., Martin, F. G., & Seehorn, D. (2019). AI4K12: Making artificial intelligence accessible to K-12 teachers and students. *Communications of the ACM, 62*(12), 26–28. https://doi.org/10.1145/3362077

Vygotsky, L. S., & Cole, M. (1978). *Mind in society: Development of higher psychological processes*. Harvard University Press.

Wei, J., Wang, X., Schuurmans, D., Bosma, M., Ichter, B., Xia, F., Chi, E., Le, Q., & Zhou, D. (2023). *Chain-of-thought prompting elicits reasoning in large language models* (No. arXiv: 2201.11903). arXiv. https://doi.org/10.48550/arXiv.2201.11903

Wood, D., Bruner, J. S., & Ross, G. (1976). The role of tutoring in problem solving. *Journal of Child Psychology and Psychiatry, 17*(2), 89–100.

Zha, S., Bragdon, M. M., Gong, N., Wang, J., Leavesley, S., Eaton, R., & Bosarge, E. (2025). A case study of integrating ai literacy education in a biology class. *International Journal of Artificial Intelligence in Education*. https://doi.org/10.1007/s40593-025-00476-8

Zhai, X., & Krajcik, J. (2024). *Uses of artificial intelligence in STEM education*. Oxford University Press.

Chapter 13
AI and the Future of Science Education: Empowering Learners? Enhancing Practice?

Grant Cooper and Kok-Sing Tang

13.1 Introduction

The emergence of artificial intelligence (AI)-including both general-purpose systems like machine learning and more recent developments such as generative AI (GenAI) has profound implications for science education. Together with related technologies such as robotics and nanotechnology, AI is increasingly seen as a catalyst for a new era of innovation poised to transform the very fabric of society (Sharma, 2023). Large language models (LLMs), a subset of GenAI, have seen widespread adoption across the globe. ChatGPT, for example, had 100 million active users each week at the time of writing (Brandl & Ellis, 2024). As AI technologies become increasingly embedded in everyday life, their capacity to reshape how we teach, learn, and engage with scientific knowledge demands close attention. We contend that the generation of scientific knowledge should no longer be seen solely as the outcome of human reasoning and empirical investigation. Instead, as we elaborate below, it ought to be understood as a collaborative endeavour between human and non-human actors, including digital technologies.

This chapter focuses specifically on the emergence of GenAI and its potential implications for the future of science education. It invites a re-examination of the learner's role in scientific inquiry. Instead of viewing students as passive recipients of AI-generated content, with the right supports, we can position them as active co-inquirers who interact with AI tools. Supporting this shift involves developing both digital literacy and epistemic humility-a recognition of both the capacities and limitations of computational reasoning (Starke et al., 2025). In this context, science

G. Cooper (✉) · K.-S. Tang
Curtin University, Perth, Australia
e-mail: grant.cooper@curtin.edu.au

K.-S. Tang
e-mail: kok-sing.tang@curtin.edu.au

© The Author(s), under exclusive license to Springer Nature Switzerland AG 2026
M. Alkış Küçükaydın (ed.), *Preparing K-12 Pre-Service Science Teachers for the Age of Artificial Intelligence*, Advances in Technology-Rich Science Education 4,
https://doi.org/10.1007/978-3-032-18030-8_13

learning is framed less as the passive absorption of static facts and more as a dynamic, continuous process of exploration, argumentation, and revision, unfolding within an ecosystem of human and artificial intelligences. This chapter proposes a forward-looking model in which learners are empowered as co-inquirers, scientific rigour is preserved, and pedagogy, curriculum, and teacher roles are reimagined. Rather than simply adapting existing practices to new technologies, the chapter advocates for a transformative approach-one that recognises GenAI not only as a tool but as a co-agent in knowledge production. In doing so, it examines both the potential benefits and challenges of GenAI, addressing the ethical, systemic, and design challenges involved in keeping science education responsive, rigorous, and meaningfully human-centred. Ultimately, this chapter presents a vision for the future of science education, one that is dynamic, inclusive, and critically informed, equipping learners to both navigate and influence a world in which scientific knowledge emerges through effective collaborations between humans and machines.

We have structured the chapter into four main sections. In the following sections, first we examine the lens of new materialism as a perspective that offers valuable insights into the nature of science and the co-agential dynamics of learning within AI-mediated environments. Subsequently, the chapter examines human–AI partnerships, highlighting the pedagogical and ethical considerations necessary for effectively integrating GenAI into science classrooms. It then discusses issues surrounding assessment practices, proposing strategies that align with the co-creative processes characteristic of GenAI-mediated learning. Ethical considerations, particularly around transparency, academic integrity, and bias, are addressed to foster responsible GenAI use. Attention is also given to the essential roles that infrastructure, policy, and professional development play in supporting the equitable and effective integration of GenAI. Finally, the chapter concludes with recommended solutions and identifies future research directions to guide educators and researchers in navigating the evolving landscape of AI-enhanced science education.

13.2 Theoretical Frame- New Materialism, Science and GenAI

As we consider the transformative potential of GenAI, it is equally important to examine the foundational assumptions that shape how knowledge is produced and how science is understood. It invites renewed attention to the notion of materiality. Materiality is central to the nature of science, yet it is often overlooked in science education research, where emphasis has typically been placed on thought and language in the formation of scientific knowledge (Hetherington et al., 2018). As Scantlebury and Milne (2019) observe, while language is vital, it "has too much power, and, with that power, there seems a concomitant loss of interest in exploring how matter contributes to both ontology and epistemology in science and science education" (p. 3). In a time increasingly shaped by tensions between the real and the artificial, it is more important than ever to root our inquiry in the material dimensions of science. Acknowledging the "parameters of artificiality, or the ways in which computers are unlike human intelligence" (Cope et al., 2021, p. 1230), and reasserting the importance of materiality, can guide us in discerning what constitutes authentic scientific knowledge and effective contemporary practices in science education.

New materialism is one perspective that offers valuable insights into the nature of science as well as the co-agential nature of learning in AI-mediated environments (Tang & Cooper, 2024). It reconceptualises science not as a purely cognitive or linguistic activity, but as an emergent, embodied practice shaped by the interactions of human and non-human agents. For our purposes, new materialism provides an effective lens through which to understand the entangled roles of learners, technologies, and material environments in scientific inquiry. It invites us to rethink what science education entails in the age of GenAI-and how knowledge itself is enacted through dynamic human–non-human relations. New materialism represents a paradigm shift in how we conceptualize the relationship between humans, matter, and knowledge (Gamble et al., 2019). Rooted in interdisciplinary traditions, drawing from philosophy, feminism, science and technology studies, and cultural theory, new materialism challenges long-standing binaries between subject and object, human and non-human, and mind and matter. New materialism foregrounds the agency of matter, presenting it as active, dynamic, and deeply entangled with human meaning-making processes.

A defining characteristic of new materialism is its departure from anthropocentrism (Sencindiver, 2017). Rather than treating humans as the sole agents of knowledge production, it posits that agency is distributed across both human and non-human actors. This is a significant reconceptualisation, especially relevant in a time where GenAI increasingly is involved in cognitive and creative processes. From this perspective, knowledge is not produced by humans in isolation. Instead, it emerges through dynamic relationships between people, technologies, tools, and the physical environment. Barad (2003) describes these relationships as '*intra-actions*', a term that highlights how entities are not independent before coming together but are actually formed through their relations. Barad's (2007) notion of *agential realism* is also central here. It suggests that matter is not a static substrate waiting to be acted upon, but rather an emergent phenomenon-something that becomes through interaction. Matter, in this sense, "is not a thing, but a doing" (Barad, 2003), and knowledge emerges not from observing matter at a distance but through participating with it. This performative view of materiality means that the properties of objects are not pre-existing; instead, they materialize through interactions with other entities, be they tools, bodies, environments, or machines.

This perspective is particularly illuminating when examining the role of GenAI in science education. Traditional educational models often place knowledge in the realm of the verbal, linguistic, and conceptual-domains well suited to GenAI's capabilities. However, as discussed, such models risk severing scientific understanding from its material roots. New materialism counters this by asserting that scientific knowledge is not solely the result of abstract reasoning or linguistic fluency, but is deeply embedded in physical practices and embodied interactions with the material world. For instance, in science, phenomena such as chemical reactions, astronomical observations, or ecological dynamics cannot be fully grasped through textual descriptions or simulations alone. Their understanding is contingent upon doing-manipulating instruments, observing outcomes, sensing changes, and constructing meaning through bodily engagement. In new materialist terms, this is where science derives its epistemic authority (Milne, 2019) not from the coherence of linguistic models, but from their alignment with material realities that challenge or affirm hypotheses through interaction.

New materialism thus encourages a rethinking of how knowledge is created, validated, and taught. It urges educators to resist the temptation of privileging symbolic abstraction-especially when augmented by GenAI-at the expense of embodied and material engagement. This has significant implications for science pedagogy. Rather than viewing GenAI as a self-sufficient source of truth, students should be guided to see it as one node in a broader epistemic network that includes experiments, artefacts, instruments, and lived experience. In sum, new materialism offers a powerful framework for navigating the challenges and opportunities presented by GenAI in science education. In a time increasingly dominated by synthetic intelligence and digital abstraction, new materialism reasserts the centrality of the physical world, not as something to be replicated or simulated, but as a co-agent in the unfolding of human understanding.

13.3 Human–AI Partnerships in Science and Science Education

A new materialist perspective accepts that the creation of scientific knowledge is not seen a linear process driven by a single agent, but as a distributed and collaborative experience between humans and machines. When used effectively, GenAI may be viewed as a co-creator-one that participates in shaping questions, scaffolding reasoning, and generating novel outputs that can extend human imagination. This framing of GenAI as a knowledge partner rather than a knowledge tool invites a deeper ontological and pedagogical inquiry: What does it mean to 'know' something in a context where some of the cognitive labour is co-produced by non-human agents? How do we ensure that learners understand both the power and limits of AI-generated insights? This paradigm shift necessitates a new way of learning and teaching science-one in which knowledge emerges from a dialogue between human, machine, and iterative critical reflection. Drawing on this perspective, both the creation and understanding of scientific knowledge can be viewed as a partnership involving both humans and GenAI, rather than attributing to either humans or GenAI alone.

Crucially, these advancements demand thoughtful ethical oversight. The very qualities that make GenAI a potentially powerful tool for learning, such as its generativity, adaptability, and rapid output, can also blur the boundaries of what constitutes reliable knowledge. In response, science educators must foreground AI literacy, encouraging students to interrogate, verify, and contextualise AI-generated content. Within the scope of this chapter, we draw on Ng et al.'s (2021) definition of AI literacy, encompassing four key elements:

- Knowing and understanding fundamental AI concepts;
- Using AI tools effectively;
- Evaluating AI outputs rigorously;
- Reflecting upon ethical implications such as fairness, transparency, and accountability.

Educators should equip students with the skills to interrogate AI-generated content, systematically verify its accuracy, and contextually interpret its implications, thereby fostering informed and ethically aware scientific practice. To Cooper et al. (2024), users... "need the acumen to question, analyse, and validate information sources, discerning the trustworthy from the misleading....(they need)... the tools and resources to cross-reference GenAI-produced content in addition to non-GenAI sourced information" (pp.115–116). Importantly, this means positioning GenAI not as a neutral or infallible tool, but as a co-agent within a broader system that shapes, constrains, and enables particular forms of knowledge. For example, students should be guided to question not only *what* an AI system says, but *why* it says it, what assumptions underlie its responses, what knowledge it draws from, and who or what may be excluded from its frame of reference.

We propose that GenAI is best understood and used as a dialogic partner in the science learning process. Rather than treating LLMs as definitive sources of information, Tang et al. (2024) frame them as dialogic agents, tools that engage learners in meaningful exchange within educational settings. This orientation is informed by Bakhtin's (1981) notion of *heteroglossia*, which emphasises the inherently dialogic nature of language, wherein every utterance weaves together multiple voices and perspectives across the continuum of communication. GenAI exemplifies heteroglossia in two ways (Tang et al., 2024). First, LLMs produce language through a synthesis of diverse data sources, forming a kind of linguistic appropriation (Bakhtin, 1981) of collective human expression. Second, interaction constitutes a co-authored process, where meaning emerges from the interplay between human input and AI-generated responses, consistent with Bakhtin's view that no utterance exists in isolation. GenAI tools have the potential to help learners formulate better questions, visualise abstract phenomena and critique underlying assumptions. Yet such capabilities also require new pedagogical understandings to guide their effective and responsible use. As GenAI becomes part of science classrooms, educators must stay critically aware of both its possibilities and constraints, ensuring it supports students' curiosity, fosters reasoning grounded in evidence, and upholds the principles of empirical inquiry. Consequently, the teacher becomes a curator of experiences and a facilitator, guiding learners to critically engage with GenAI outputs while grounding their understanding in empirical evidence and conceptual rigour. Students, in turn, are invited to become co-investigators in this learning environment.

The goal is not to produce students who are merely adept at using AI tools, but rather to nurture scientifically literate citizens who can engage with emerging technologies critically and creatively. This entails an expanded view of inquiry-one that embraces ambiguity, values questioning over certainty, and recognises that meaning-making in science now occurs across a broader network of human and non-human actors. GenAI, when thoughtfully integrated, can be a powerful ally in this endeavour, not by simplifying science but by deepening students' engagement with its complexity. GenAI may seduce users into accepting its outputs as authoritative, thereby obscuring the essential role that materiality plays in grounding scientific claims. To mitigate this, science educators must ensure that learners are not merely adept at prompting or evaluating GenAI outputs, but are deeply attuned to the differences in epistemic authority between human and AI-generated knowledge. In the face of misinformation, algorithmic bias, and the blackbox opacity of LLMs, the stakes are high. Learners must be trained not only to ask whether an AI-generated output is plausible, but whether it is *true*-and how such truth can be tested, traced, and substantiated through material engagement with the world.

As mentioned, the remainder of the chapter examines key issues surrounding assessment practices, proposing strategies aligned with the co-creative nature of GenAI-mediated learning. Subsequently, ethical considerations, particularly transparency, academic integrity, and bias, are explored. We also look at the importance of infrastructure, policy frameworks, and professional development in supporting equitable and effective GenAI integration. Finally, the chapter concludes by recommending practical solutions and outlining future research directions, guiding educators and researchers in navigating the dynamic landscape of AI-enhanced science education.

13.4 Assessment

The integration of GenAI into science education compels a reimagining of assessment practices. Traditional models of assessment, emphasising the recall of scientific facts, individual authorship, and standardised outputs, are increasingly misaligned with the hybrid human–machine processes that now characterise learning environments embedded with GenAI.

Assessing co-created knowledge involves attending to the ways in which learners interact with GenAI throughout their inquiry. Rather than treating AI-generated content as suspect or inauthentic, science educators can design assessments that foreground students' ability to interrogate, modify, and justify their use of GenAI in the context of a material reality. For example, students could be asked to record their prompt strategies, explain the reasoning behind any edits or rejections of GenAI-generated responses, and critically assess specific outputs in relation to their scientific goals and the results of hands-on investigations, such as experiments or observations. The evolving role of GenAI also raises new challenges around academic integrity. As AI tools become more fluent and capable, the distinction between human-authored and machine-generated content grows increasingly blurred. Rather than relying solely on punitive models and plagiarism detection software, science educators have an opportunity to embed critical AI literacy into marking criteria, rewarding students not only for the epistemic soundness of their claims but also for their ability to question, validate, and contextualize AI outputs with care and insight.

Material engagement remains central to scientific learning and should continue to serve as an anchor for assessment. The use of GenAI should not replace opportunities for direct, hands-on scientific investigation. In envisioning future-facing forms of assessment, it is important to consider tasks that reflect the competencies students will require in a GenAI-mediated world. Scenario-based projects, peer feedback on GenAI inquiries, and reflective writing on ethical dilemmas around algorithmic bias or misinformation are all examples of authentic assessment strategies that could support this goal. Such tasks encourage students to move beyond the passive consumption of AI-generated content toward active critique, collaboration, and meaning-making. They also provide opportunities for teachers to assess qualities like epistemic humility, curiosity, and critical thinking, attributes that are difficult to measure through standardized testing but are essential in cultivating scientifically literate citizens.

13.5 Ethical Considerations

The integration of GenAI in science education introduces ethical considerations that educators must navigate carefully. To maintain rigour and trustworthiness, students need to learn how to effectively disclose how GenAI was used in their work, including the specific tools engaged, their purpose, and the nature of human oversight or modification. These steps promote the ethical use of GenAI and help students metacognitively reflect on their learning process. Transparent disclosure helps promote academic integrity and encourages reflective practice. In keeping with our commitment to transparency, and as an example of the type of openness we encourage from students, this is how we have used GenAI in the development of this paper:

- The GenAI application and version (e.g., ChatGPT 4o)
- The period of use (March to July 2025)
- Justification (idea generation, including suggestions for paper structure and phrasing, as well as to enhance the clarity and readability of sections)
- Prompts/input (examples: ChatGPT prompt: rewrite: [paste sentence], Suggest ways to structure a section on GenAI, science education and new materialism)

- Authors' review and acknowledgment of responsibility: (e.g., We acknowledge that the authors have carefully reviewed and edited the manuscript and take full responsibility for the text)

As discussed, GenAI is prone to hallucinations-outputs that sound plausible but are factually incorrect. In science education, such errors can have serious consequences, especially when students lack the expertise to detect them. Teachers must model critical engagement, routinely fact-check GenAI outputs, and encourage students to triangulate information from multiple sources.

13.6 Representation, Bias and Critical Literacies

Another significant concern is the inherent bias in the data that GenAI models are trained on, which can reinforce biases related to race, gender, and socio-economic status (SES) (Zhai, 2022). GenAI's outputs often carry embedded biases, hallucinations, or unexamined assumptions that may mislead uncritical users or distort scientific understanding (Cooper, 2023). Although OpenAI has not disclosed the full details of ChatGPT's training data, it is widely acknowledged that the model is primarily optimized for English. This reflects not only the linguistic composition of the data but also reinforces English as the dominant language of interaction and information exchange (OpenAI, 2023). Such optimization enhances the model's ability to capture linguistic nuance in English but also entrenches epistemic bias. In science education, this has significant implications for knowledge representation, particularly for students and educators working in non-English contexts. Moreover, OpenAI concedes that ChatGPT exhibits a bias towards Western perspectives, and that many of its safeguards for harmful content have only been tested in English (OpenAI, 2023). This linguistic and cultural imbalance means that science-related content generated by the model may prioritise Western framings of science, potentially marginalizing Indigenous knowledge systems, alternative worldviews, and localized scientific perspectives. From a new materialist standpoint, these biases are not merely discursive, they are materially instantiated in the infrastructures, datasets, and feedback loops through which GenAI systems operate. The training data, algorithmic structures, and user interactions that shape GenAI outputs are themselves material agents that participate in the co-production of knowledge, influencing what becomes visible, sayable, and valued in science classrooms. In addition, the bias is reinforced through patterns of user interaction. As Krist and Kubsch (2023) point out, ChatGPT's continual learning from user feedback enables it to become more proficient in the languages and sociocultural norms of frequently represented groups, particularly English-speaking users from affluent countries. For science education, the consequences are twofold: first, learners engaging with GenAI tools like ChatGPT may unknowingly be exposed to a narrow conception of scientific knowledge, predominantly rooted in Western epistemology; second, students from underrepresented linguistic or cultural backgrounds may find their perspectives under acknowledged or misrepresented. These dynamics risk eroding cultural relevance and pedagogical inclusivity, posing a significant barrier to cultivating a kind of scientific literacy that honours diverse languages and perspectives.

Moreover, Cooper and Tang (2024) found that GenAI-generated classroom images often reproduced stereotypical representations of science educators and learning environments (e.g. white lab coats, beakers, teacher-centered learning environments). Despite gender and ethnic diversity, these visual outputs often reflect a narrow conception of what science looks like-and who gets to participate. These visual cues, when uncritically accepted, may alienate students who do not see themselves reflected in dominant scientific narratives. To counter this, science educators should treat GenAI-generated content not as neutral artefacts, but as texts for critical reading. Classroom practices might include comparing AI-generated images with real-world lab settings, analyzing who is included or excluded in visual representations, or asking students to generate their own inclusive scientific imagery. Moreover, this critical engagement must extend beyond image analysis. Students should be encouraged to reflect on how other forms of visual representations, such as AI-generated videos circulated, influence their perceptions of scientific identity and legitimacy. By deconstructing the semiotics of GenAI outputs. Integrating these reflective practices into science lessons not only fosters inclusivity but also equips students with the critical literacy skills essential for navigating a world saturated with generated content.

13.7 Infrastrure, Policy and Professional Development

While some educational institutions are well-equipped with bandwidth, devices, and institutional support, many others, especially those in underserved or marginalized communities, often lack the essential infrastructure to engage meaningfully with emerging technologies such as GenAI (Hendawy, 2024). Such technological disparity threatens to deepen existing inequities in science education by granting some students privileged access, while others remain excluded. Addressing this requires targeted investment in digital infrastructure, particularly in terms of device access, internet connectivity, and secure systems for managing student data. From a policy perspective, educational leaders should provide science educators with clearer guidance on the pedagogical use of GenAI. Current policy frameworks lag behind the rapid pace of technological development, likely leaving many science teachers uncertain about how to integrate GenAI tools responsibly and effectively. Beyond general digital competence, there is a need to explicitly incorporate AI literacy as both a set of tools to use and a topic for critical inquiry into science learning. Professional development is essential for supporting science teachers to use GenAI in ways that are pedagogically effective. Exposure to GenAI tools without ongoing, context-relevant support is unlikely to lead to meaningful pedagogical innovation. Instead, science educators need structured opportunities for collaborative inquiry into how GenAI might support authentic scientific learning. These professional learning experiences should be iterative, reflective, and embedded in communities of practice that encourage innovation while attending to the responsible use of GenAI.

13.8 Solutions and Recommendations

In light of the issues raised surrounding human-AI partnerships highlighting the pedagogical and ethical uses of GenAI, Table 13.1 summarises the recommended solutions to guide educators and researchers in using GenAI within science education.

13.9 Future Research Directions

Future research should explore how learners and teachers engage with GenAI as co-partners in knowledge production. More research is needed to examine how GenAI tools influence students' epistemic beliefs, conceptual understanding, and scientific reasoning when embedded in science education experiences. There is also a pressing need to promote critical AI literacy understandings that help students navigate GenAI bias, assess the credibility of AI outputs, and reflect on ethical and legal implications such as authorship, privacy, and academic integrity. Additionally, future work should investigate how the roles and professional identities of teachers are evolving in response to the use of GenAI in science classrooms.

13.10 Conclusion

GenAI presents both opportunity and challenges in science education. When embedded effectively, it can transform classrooms into learning environments where human insight and artificial creativity coalesce. It invites a science pedagogy that is exploratory, collaborative, embodied, and ethically attuned. The advent of GenAI invites science educators to reconsider not just how science is taught, but what it means to know, inquire, and create in a world co-shaped by artificial intelligence. Rather than resisting or uncritically embracing these tools, this chapter has argued for a critical, inclusive, and materially grounded engagement with GenAI in science education. Drawing on new materialist theory, we have suggested that knowledge emerges through the entangled intra-actions of humans, technologies, and material environments. GenAI, when thoughtfully integrated, has the potential to deepen scientific inquiry and support co-creative learning experiences. However, this promise comes with significant pedagogical considerations. As we move forward, science educators must be supported through professional development, guided by effective policy, and empowered to foster critical AI literacy among

Table 13.1 Summary of when, why, and how to integrate GenAI into science education

When	Why	How
When introducing GenAI in science learning	To support AI literacy	• Facilitate dialogic inquiry using GenAI as a co-agent • Encourage guided prompt experimentation • Pair with teacher-led clarification and concept mapping
When engaging in scientific modelling, simulation, or prediction	To support conceptual abstraction	• Use GenAI to model systems (e.g. food webs, chemical reactions) • Compare AI outputs with textbook examples or experimental data • Use simulations as springboards for empirical testing
When cultivating epistemic humility and critical AI literacy	To challenge over-reliance on GenAI	• Analyse GenAI hallucinations and bias • Compare AI outputs to empirical and peer-reviewed sources • Discuss how scientific knowledge is validated through observation and repetition
When promoting inclusive and culturally responsive science	To critique dominant AI representations and expand narratives of science	• Interrogate AI-generated images, videos, and texts for bias • Encourage students to create inclusive alternative representations using GenAI • Compare GenAI outputs with diverse real-world examples
When supporting co-inquiry and student agency	To position learners as co-creators rather than passive consumers	• Invite students to test, revise, and iterate on AI-generated ideas • Encourage documentation of learning choices and decision-making processes
When building ethical, legal, and responsible practices	To model transparency, privacy, and ethical AI use	• Require students to disclose AI usage in their submissions • Include prompt logs or reflection sections in assessments •Discuss intellectual property/ data privacy

their students. In embracing the potential of GenAI, we must also remain attuned to what grounds science, such as material evidence, empirical integrity, and the pursuit of understanding through dialogue, experimentation, and reflection.

References

Bakhtin, M. M. (1981). *The dialogic imagination: Four essays*. University of Texas Press.

Barad, K. (2003). Posthumanist performativity: Toward an understanding of how matter comes to matter. *Signs: Journal of Women in Culture and Society, 28*(3), 801–831.

Barad, K. (2007). *Meeting the universe halfway: Quantum physics and the entanglement of matter and meaning*. The University of Chicago Press

Brandl, R., & Ellis, C. (2024). *Chatgpt statistics and user numbers 2024—Openai chatbot*. Tooltester. https://www.tooltester.com/en/blog/chatgpt-statistics/

Cooper, G. (2023). Examining science education in ChatGPT: An exploratory study of generative artificial intelligence. *Journal of Science Education and Technology, 32*(3), 444–452.

Cooper, G., & Tang, K. S. (2024). Pixels and pedagogy: Examining science education imagery by generative artificial intelligence. *Journal of Science Education and Technology*. Advance online publication. https://doi.org/10.1007/s10956-024-10104-0

Cooper, G., Tang, K.S, & Rappa, N. (2024). *Generative Artificial Intelligence as Epistemic Authority?: Perspectives from Higher Education*. In H. Crompton & D. Burke, D. (Eds.). *Artificial intelligence applications in higher education: Theories, ethics, and case studies for universities* (1st ed.). Routledge. https://doi.org/10.4324/9781003440178

Cope, B., Kalantzis, M., & Searsmith, D. (2021). Artificial intelligence for education: Knowledge and its assessment in AI-enabled learning ecologies. *Educational Philosophy and Theory, 53*(12), 1229–1245. https://doi.org/10.1080/00131857.2020.1728732

Gamble, C. N., Hanan, J. S., & Nail, T. (2019). What is new materialism? *Angelaki, 24*(6), 111–134.

Hendawy, M. (2024). *The intensified digital divide: Comprehending GenAI*. Internet Policy Review. https://policyreview.info/articles/news/intensified-digital-divide-comprehending-genai/1772

Hetherington, L., Hardman, M., Noakes, J., & Wegerif, R. (2018). Making the case for a material-dialogic approach to science education. *Studies in Science Education, 54*(2), 141–176.

Krist, C., & Kubsch, M. (2023). Bias, bias everywhere: A response to Li et al. and Zhai and Nehm. *Journal of Research in Science Teaching, 60*(10), 2395–2399. https://doi.org/10.1002/tea.21913

Milne, C. (2019). The materiality of scientific instruments and why it might matter to science education. *Material practice and materiality: Too long ignored in science education*, 9–23. *Education* (pp. 9–25). Springer. 10.1007/978-3-030-01974-7_1

Ng, D. T. K., Leung, J. K. L., Chu, S. K. W., & Qiao, M. S. (2021). Conceptualizing AI literacy: An exploratory review. *Computers and Education: Artificial Intelligence, 2*, Article 100041.

OpenAI. (2023). *Educator considerations for ChatGPT*. https://platform.openai.com/docs/chatgpt-education

Sencindiver, S. Y. (2017). *New materialism*. Literary and critical theory. Oxford University Press.

Sharma, R. (2023). The transformative power of AI as future GPTs in propelling society into a new era of advancement. *IEEE Engineering Management Review, 51*(4), 215. https://doi.org/10.1109/EMR.2023.10342241

Starke, G., Sobieska, A., Knochel, K., & Buyx, A. (2025). Epistemic humility meets virtual reality: Teaching an old ideal with novel tools. *Journal of Medical Ethics*.

Tang, K.-S., & Cooper, G. (2024). The role of materiality in an era of generative artificial intelligence. *Science & Education, 34*(2), 731–746. https://doi.org/10.1007/s11191-024-00508-0

Tang, K.-S., Cooper, G., Rappa, N., Cooper, M., Sims, C., & Nonis, K. (2024). A dialogic approach to transform teaching, learning & assessment with generative AI in secondary education: A proof of concept. *Pedagogies: An International Journal, 19*(3), 493–503. https://doi.org/10.1080/1554480X.2024.2379774

Zhai, X. (2022). Artificial intelligence in science education: A systematic review. *Journal of Science Education and Technology, 31*(2), 203–215. https://doi.org/10.1007/s10956-021-09924-3

Correction to: Preparing K-12 Pre-Service Science Teachers for the Age of Artificial Intelligence

Menşure Alkış Küçükaydın

Correction to: Chapters 9 and 12 in M. Alkış Küçükaydın (ed.), *Preparing K-12 Pre-Service Science Teachers for the Age of Artificial Intelligence*, Advances in Technology-Rich Science Education 4, https://doi.org/10.1007/978-3-032-18030-8

In the original version of this book, two authors have been removed in Chapter 12 and a spelling mistake in the title has been corrected in Chapter 9. The book has been updated with the changes.

The updated versions of these chapters can be found at
https://doi.org/10.1007/978-3-032-18030-8_9
https://doi.org/10.1007/978-3-032-18030-8_12

© The Author(s), under exclusive license to Springer Nature Switzerland AG 2026
M. Alkış Küçükaydın (ed.), *Preparing K-12 Pre-Service Science Teachers for the Age of Artificial Intelligence*, Advances in Technology-Rich Science Education 4,
https://doi.org/10.1007/978-3-032-18030-8_14

Conclusion

The chapters presented in this book address the integration of AI into science education not merely as a technological innovation but also as a pedagogical, ethical, and epistemological transformation. Prospective teachers' interactions with AI yield multidimensional learning outcomes, including conceptual coherence, higher-order thinking skills, emotional resilience, and ethical awareness. In this context, models such as AI-TPACK, CILP, and Design Thinking provide theoretical tools to explain how teachers make sense of AI and integrate it into the classroom.

Three fundamental orientations must be adopted for AI to be used effectively and ethically in science education:

- Conceptual integrity: AI tools should support students' understanding of scientific concepts holistically, not piecemeal. This is possible not only by presenting information but also by triggering processes of inquiry, modeling, and synthesis.
- Ethical orientation: Issues such as data privacy, algorithmic bias, academic integrity, and equal access should not be overlooked in the classroom use of AI. Teachers should have ethical reflexes that enable them to provide guidance on these issues.
- Teacher agency: The opportunities offered by AI should empower rather than weaken the role of the teacher. Teachers should be positioned as actors who direct AI for pedagogical purposes, monitor students' emotional and cognitive processes, and structure the learning environment in accordance with ethical principles.

Based on the book's findings, the following recommendations are made to restructure science education in the AI era in a more inclusive, questioning, and ethically grounded manner:

1. Priority should be given to ai literacy inc teacher education programs: The working logic, limitations, potential biases, and pedagogical use of AI systems should be systematically taught to pre-service teachers. These contents should be

© The Editor(s) (if applicable) and The Author(s), under exclusive license to Springer Nature Switzerland AG 2026

M. Alkış Küçükaydın (ed.), *Preparing K-12 Pre-Service Science Teachers for the Age of Artificial Intelligence*, Advances in Technology-Rich Science Education 4,
https://doi.org/10.1007/978-3-032-18030-8

integrated into instructional technology courses and supported by applications such as microteaching, scenario analysis, and reflective portfolios.

2. AI-supported inquiry-based learning environments should be developed: Students should be encouraged to formulate hypotheses, analyze data, develop conceptual models, and discuss their results with AI. Teacher guidance, management of epistemic emotions, and ethical awareness should be prioritized in these processes.
3. AI integration should be evaluated using UDL and SAMR frameworks: AI tools should be evaluated in terms of accessibility, diversity of representation, and freedom of expression; adaptable solutions should be offered for students with special needs; and technology should be used not merely as a replacement but as a means of redefinition.
4. Ethical decision-making skills should be central to teacher education: The classroom use of AI should be addressed alongside issues such as data security, academic integrity, algorithmic bias, and digital equity; pre-service teachers should develop critical reflection on these topics.
5. Cross-cultural comparative research should be encouraged: Studies examining how AI is experienced in science classrooms across different countries, how pedagogical approaches vary, and how students' epistemic emotions are managed are important for establishing the balance between universality and local context in teacher education programs.
6. AI must be positioned as a tool that empowers teachers: AI should be seen not as a replacement for teachers, but as an assistant that supports teachers' pedagogical decisions, facilitates the monitoring of students' learning processes, and provides conceptual guidance.

Index

© The Editor(s) (if applicable) and The Author(s), under exclusive license to Springer Nature Switzerland AG 2026

M. Alkış Küçükaydın (ed.), *Preparing K-12 Pre-Service Science Teachers for the Age of Artificial Intelligence*, Advances in Technology-Rich Science Education 4, https://doi.org/10.1007/978-3-032-18030-8

FSC
www.fsc.org
MIX
Papier | Fördert
gute Waldnutzung
FSC® C083411

Kolibri 360 GmbH
Ferdinand-Jühlke-Straße 7
99095 Erfurt, Deutschland
produktsicherheit@kolibri360.de